AF746481

AUTOMATED DATA RETRIEVAL IN ASTRONOMY

ASTROPHYSICS AND SPACE SCIENCE LIBRARY

A SERIES OF BOOKS ON THE RECENT DEVELOPMENTS
OF SPACE SCIENCE AND OF GENERAL GEOPHYSICS AND ASTROPHYSICS
PUBLISHED IN CONNECTION WITH THE JOURNAL
SPACE SCIENCE REVIEWS

Editorial Board

J. E. BLAMONT, *Laboratoire d'Aeronomie, Verrières, France*

R. L. F. BOYD, *University College, London, England*

L. GOLDBERG, *Kitt Peak National Observatory, Tucson, Ariz., U.S.A.*

C. DE JAGER, *University of Utrecht, The Netherlands*

Z. KOPAL, *University of Manchester, England*

G. H. LUDWIG, *NOAA, National Environmental Satellite Service, Suitland, Md., U.S.A.*

R. LÜST, *President Max-Planck-Gesellschaft zur Förderung der Wissenschaften, München, F.R.G.*

B. M. McCORMAC, *Lockheed Palo Alto Research Laboratory, Palo Alto, Calif., U.S.A.*

H. E. NEWELL, *Alexandria, Va., U.S.A.*

L. I. SEDOV, *Academy of Sciences of the U.S.S.R., Moscow, U.S.S.R.*

Z. ŠVESTKA, *University of Utrecht, The Netherlands*

VOLUME 97
PROCEEDINGS

AUTOMATED DATA RETRIEVAL IN ASTRONOMY

PROCEEDINGS OF THE 64th COLLOQUIUM
OF THE INTERNATIONAL ASTRONOMICAL UNION
HELD IN STRASBOURG, FRANCE, JULY 7–10, 1981

Edited by

C. JASCHEK
Observatoire de Strasbourg, France

and

W. HEINTZ
Swarthmore College, Swartmore, U.S.A.

D. REIDEL PUBLISHING COMPANY
DORDRECHT : HOLLAND / BOSTON : U.S.A.
LONDON : ENGLAND

Library of Congress Cataloging in Publication Data

CIP

International Astronomical Union. Colloquium (64th : 1981 :
Strasbourg, France)
Automated data retrieval in astronomy.

(Astrophysics and space science library ; v. 97)
Includes index.
1. Astronomy–Data processing–Congresses. 2. Information storage and retrieval systems–Astronomy–Congresses. 3. Astronomy–Information services–Congresses. I. Jaschek, Carlos. II. Heintz, W. III. Titles. IV. Series.
QB51.3.E43I57 1981 522'.028'54 82-7676
ISBN 90-277-1435-5 AACR2

UNIVERSITY COLLEGE
SCIENCE
LIBRARY
CARDIFF

Published by D. Reidel Publishing Company,
P.O. Box 17, 3300 AA Dordrecht, Holland.

Sold and distributed in the U.S.A. and Canada
by Kluwer Boston Inc.,
190 Old Derby Street, Hingham, MA 02043, U.S.A.

In all other countries, sold and distributed
by Kluwer Academic Publishers Group,
P.O. Box 322, 3300 AH Dordrecht, Holland.

D. Reidel Publishing Company is a member of the Kluwer Group.

All Rights Reserved
Copyright © 1982 by D. Reidel Publishing Company, Dordrecht, Holland
No part of the material protected by this copyright notice may be reproduced or utilized in any form or by any means, electronic or mechanical including photocopying, recording or by any informational storage and retrieval system, without written permission from the copyright owner

Printed in The Netherlands

TABLE OF CONTENTS

G. EDITORIAL POLICIES AND NOMENCLATURE

H. DATA IN ASTRONOMY

I. DATA IN SPACE ASTRONOMY

PREFACE

The idea of this Colloquium came during the XVIIth General Assembly of the I.A.U. at Montreal. The meeting was organized under the auspices of I.A.U. Commission 5 (Documentation and Astronomical Data). The Scientific Organizing Committee consisted of C. Jaschek (chairperson), O. Dluzhnevskaya, B. Hauck (vice chairperson), W. Heintz, P. Lantos, Th. Lederle, J. Mead, G. Ruben, Y. Terashita, G. Wilkins.

The members of this Committee are to be thanked for their devotion to the organization of what turned out to be a very successful meeting. The program was organized so as to cover most of the aspects concerning work with machine readable data. In a certain sense it is the development of the subjects of I.A.U. Colloquium 35 "Compilation, critical evaluation and distribution of stellar data" held at Strasbourg in 1976. The meeting was opened by welcoming addresses delivered by Dr A. Florsch, Director of the Strasbourg Observatory, Prof. H. Curien, President of the European Science Foundation and Prof. W. Heintz, President of I.A.U. Commission 5. The sessions were devoted to the following subjects : Existing data centers, Data networks, New hardware, Recent software developments, Bibliographical services, Copyright, Editorial policies and nomenclature, Data in astronomy and Data in space astronomy. The different sessions were chaired by G.A. Wilkins, J. Mead, S. Lavrov, W. Heintz, P. Lantos, M. McCarthy, J. Delhaye and G. Westerhout.

On July 9, Dr A. Florsch spoke on the occasion of the Centenary of the Strasbourg Observatory and described briefly the various activities of the institution over the past century.

The meeting closed with a general discussion and the final recommendations by Prof. B. Hauck.

The local organization was handled very efficiently by the Local Organizing Committee composed of A. Florsch (chairperson), Ch. Bruneau, D. Egret, A. Fresneau, C. Jaschek and F. Ochsenbein. The work of the Committee was greatly helped by the collaboration of a number of volunteers from the Strasbourg Observatory.

The Colloquium was attended by ninety scientists from fifteen countries.

Financial support was received from the International Astronomi-

C. Jaschek and W. Heintz (eds.), Automated Data Retrieval in Astronomy, xi–xii.
Copyright © 1982 by D. Reidel Publishing Company.

cal Union, the Institut National d'Astronomie et Géophysique, the Ministère des Universités, the Observatoire de Strasbourg and the Municipalité de Strasbourg.

The meeting took place at the Institute Le Bel of the Louis Pasteur University of Strasbourg. We thank all these institutions for their support.

C. JASCHEK

January 1982 W. HEINTZ

SCIENTIFIC ORGANIZING COMMITTEE

O. Dluzhnevskaya, B. Hauck (vice chairperson), W. Heintz, C. Jaschek (Chairperson), P. Lantos, Th. Lederle, J. Mead, G. Ruben, Y. Terashita, G. Wilkins.

LOCAL ORGANIZING COMMITTEE

Ch. Bruneau, D. Egret, A. Florsch (Chairperson), A. Fresneau, C. Jaschek, F. Ochsenbein.

LIST OF PARTICIPANTS

ACKER A., Observatoire de Strasbourg, France
ALBRECHT R., Institute for Astronomy, Vienna, Austria
ARGUE N., The Observatories, Cambridge,U.K.
BARBIER M., Observatoire de Marseille, France
BENACCHIO L., Osservatorio di Padova, Italy
BERARDINI Cl., Observatoire de Meudon, France
BERGWALL M., Uppsala Observatory, Sweden
BIJAOUI A., Observatoire de Nice, France
BISCHOFF M., Observatoire de Strasbourg, France
BONOLI C., Osservatorio di Padova, Italy
CARNOCHAN D., University College, London, U.K.
COLLINS M., INSPEC, Hitchin, U.K.
CURIEN H., C.N.E.S., Paris, France
DAVIS R.J., Smithsonian Astrophysic Observatory, Cambridge, U.S.
DELHAYE J., Observatoire de Paris, France
DE PABLO D., E.S.A., Villafranca, Spain
DICKEL H.R., Illinois Observatory, Urbana, U.S.
DLUZHNEVSKAYA O., Astronomical Council, Moscow, U.S.S.R.
DUBOIS P., Observatoire de Strasbourg, France
DUFLOT M., Observatoire de Marseille, France
EBEL G., Fachinformationszentrum, Karlsruhe, Germany
EGRET D., Observatoire de Strasbourg, France
EICHHORN H., University of Florida, Gainesville, U.S.
FLORSCH A., Observatoire de Strasbourg, France
FERNANDEZ A., Observatoire de Meudon, France
FRESNEAU A., Observatoire de Strasbourg, France
GLIESE W., Astronomisches Rechen Institut, Heidelberg, Germany
GROSBOL P., E.S.O., Garching, München, Germany
GUERASSIMOV E., U.N.E.S.C.O., Paris, France
HAUCK B., Lausanne Observatory, Sauverny, Switzerland
HECK A., E.S.A., Villafranca, Spain
HEINTZ W., Swarthmore College, Swarthmore, U.S.A.
HERING R., Astronomisches Rechen Institut, Heidelberg, Germany
JAHREISS H., Astronomisches Rechen Institut, Heidelberg, Germany
JANES K., Boston Observatory, Boston, U.S.A.
JASCHEK C., Observatoire de Strasbourg, France
JASCHEK M., Observatoire de Strasbourg, France
KALINKOV M., Academy of Sciences, Sofia, Bulgaria
KOGOSHVILI N., Abastumani Observatory, U.S.S.R.
KRISTIAN J., Mt Wilson Observatory, Pasadena, U.S.A.
LAHMEK R., Observatoire de Meudon, France

LALOE S., Institut d'Astrophysique, Paris, France
LANTOS P., Observatoire de Meudon, France
LAVROV S., Inst. Theoretical Astr., Leningrad, U.S.S.R.
LEQUEUX J., Observatoire de Meudon, France
LOCKE M.C., N.A.S.A., Greenbelt, U.S.A.
LYNGA G., Lund Observatory, Sweden
McCARTHY M., Specola Vaticana, Vatican
MARCOUT J., Observatoire de Strasbourg, France
MISTRIK I., G. 10, Heidelberg, Germany
MEAD J., N.A.S.A., Greenbelt, U.S.A.
NICOLET B., Observatoire de Genève, Sauverny, Switzerland
NISHIMURA S., Tokyo Astronomical Observatory, Mitaka, Japan
NORTH P., Observatoire de Lausanne, Sauverny, Switzerland
NOTTALE L., Observatoire de Meudon, France
OCHSENBEIN F., Observatoire de Strasbourg, France
OJA T., Kvistaberg Observatory, Sweden
PATUREL G., Observatoire de Lyon, France
PERRYMAN M., E.S.T.E.C. Noordwijk, Netherlands
POLECHOVA P., Praha Observatory, Czechoslovakia
RUSCONI L.S., Osservatorio di Trieste, Italy
SALLE Y., EURONET, Luxembourg, Luxemburg
SANTIN P., Osservatorio di Trieste, Italy
SCHEIDECKER J.P., Observatoire de Nice, France
SCHIMINOVICH S., American Institut of Physics, New York, U.S.A.
SCHMADEL L., Astronomisches Rechen Institut, Heidelberg, Germany
SCHMITZ M., N.A.S.A., Greenbelt, U.S.A.
SCHUTZ A., E.S.O.C., Darmstadt, Germany
SEDMAK G., Osservatorio di Trieste, Italy
SIGNORE M., Observatoire de Meudon, France
SIMON P., Observatoire de Meudon, France
SPITE F., Observatoire de Meudon, France
STEIN W.L., Dahlgreen Lab., Dahlgreen, U.S.A.
STENHOLM B., Lund Observatory, Sweden
STRAIZYS V., Vilnius Observatory, U.S.S.R.
TERASHITA Y., Kanazawa Institute, Ishikawa, Japan
TERLOUW J.P., Kapteyn Astr. Lab., Groningen, Netherlands
TRITTON S., Royal Observatory of Edinburgh, U.K.
TURON C., Observatoire de Meudon, France
UPGREN A., van Vleck Observatory, Middleton, U.S.A.
VALBOUSQUET A., Observatoire de Strasbourg, France
VETTE J.I., N.A.S.A., Greenbelt, U.S.A.
WAGNER R.L., Ford A.I.C.C., Colorado Springs, U.S.A.
WARREN W., N.A.S.A. Greenbelt, U.S.A.
WENGER M., Observatoire de Strasbourg, France
WESTERHOUT G., Naval Observatory, Washington D.C., U.S.A.
WILKINS G.A., Royal Greenwich Observatory, Herstmonceux, U.K.
WOOD R., Royal Greenwich Observatory, Herstmonceux, U.K.
WORLEY Ch., Naval Observatory, Washington D.C., U.S.A.

HERITAGE ON TAPE (Inaugural address)

Wulff D. Heintz
Dept. of Astronomy, Swarthmore College

For your friendly welcome, Dr.Florsch and Dr.Curien, please accept our sincere thanks. The astronomers specialising in data documentation, and their Commission in the IAU, are very grateful indeed to be able to meet here again after our first conference in 1976. It is also an honor that this Colloquium with participation from 15 countries is leading off a series of events celebrating the centennial of the Strasbourg Observatory. The addition of astronomy to your academic community has been one of continuing success, characterised by a long list of important publications, and of well-known names associated with the place through the decades. Thus you possess an active center of research and documentation in astronomy, and this in a charming, hospitable city of extraordinary beauty. Certainly it is to be ascribed to the attraction of the place and to the renown of our hosts as well as to the important subject of the Colloquium, that your invitation has drawn so large a response.

Between the generation and the use of our products lie publishing, abstracting and tagging, collection, intermediary or selective evaluation, repackaging, and dissemination or marketing. This quotation does not come from an economics textbook on merchandise, but from a CODATA publication. Our research experience has acquainted us not only with the production and application of data, the ends of the chain, but also with the information brokerage steps in between, which simply are inevitable once a society and its supply/demand transaction rate reach a certain size. This applies to supermarket sales as well as to star positions. Astronomy with its long history, and with the longevity of gathered data has early become aware of the problem, and it was already in 1921, in preparation for the first IAU assembly, that M.Baillaud of the Paris Observatory was invited to form a Commission which - first of all at that time - was to concern itself with bibliography and abstracting. Thus we have reached, along with the Strasbourg Observatory centennial, the 60th anniversary of Commission 5. We have witnessed the strong additional emphasis on data documentation, storage, and processing in the last 15 years, when computer systems had become available to do large-scale jobs in that area, and we have met here five years ago to develop

C. Jaschek and W. Heintz (eds.), Automated Data Retrieval in Astronomy, xvii–xx.
Copyright © 1982 by D. Reidel Publishing Company.

ideas on what can be, and ought to be, done by way of coordination. The technical development since has been so breathtakingly fast, the scientific personnel involved also increasing, that another conference seemed desirable, and its initiator, C.Jaschek, needed little persuasion to get it approved.

Much has happened in the meantime, as can be seen from the data center bulletins and other reports. Machine-readable holdings have substantially increased. The spreading of computer terminals linking with databanks and of networks between centers puts a much larger clientele into reach of the data which they need, and can now obtain wholesale. B.Hauck reported on this contribution from astronomy at the last CODATA meeting. The 75-page listing of existing facilities by Jaschek at our preceding conference would now certainly be thicker. Important progress has been made toward an object nomenclature and a controlled vocabulary for the purpose of data tagging with good reliability of completeness. We have at least the promise of various IAU Commissions to try and promote such standards, and it remains to convince also the publishers of primary literature - most of which is not in astronomical hands - where sometimes vague and inconsistent publishing policies can be sources of mistakes. Toward this goal the efforts toward standards for presentation of data in primary journals along CODATA guidelines should also prove helpful.

Scientists in Physics, Chemistry, and Mineralogy, for instance, require all kinds of properties of a million compounds and minerals to be contained in their databanks. Most of such data are time-invariant and reproducible, making revision and augmentation - along with an elaborate multi-purpose access - the main tasks for the compiler. On the other hand, geosciences have been flooded - particularly through satellites - with huge amounts of instantaneous data. One weather satellite yields 10^7 binary digits per transmission, or something like 10^{12} bits of output per year, and the synoptic readings, on which one weather analysis is based, contain 10^6 to 10^7 bits with a required compiling/processing time under 1 hour through the World Meteorological Network. But many of these data are of transient significance only, and need be stored merely in condensed form for long-term reference.

Astronomy has an in-between position, facing various requirements for its various needs. Large data material is (or may be assumed to be) time-dependent and not repeatable, and the interval between acqusition and perusal frequently is long. We are now in a position that large enterprises like the Palomar Survey or the 21-cm mapping could soon be repeated and automatically compared with first-epoch tapes for variations. Given also the interdisciplinary nature of much of the research, particularly in Astro- and Space Physics, we should provide for a data tagging system with as much compatibility as we can afford without sacrificing those classification features which we - although a smaller community than Physics - deem indispensable for our research approach. And finally, since Astronomy has always been a struggle with noise levels, caveats on the degree of reliability of data need special attention. In older observations the lower precision may outweigh the longer timespan.

A related issue recently caught my attention, namely the guidelines for abstracts. They could be improved in some ways: homogeneity, emphasis on data flags and on inexpensive, limited-character reproduction by abstracting services, and enforcement by the primary periodicals.

To those who instruct students and advise younger colleagues, a suggestion may be addressed: Make them aware of the means - bibliographic and data storage - that are at their disposal. Part of the scientific community does have a traditionalist attitude toward new mechanisms in information, the shelves of printed books and catalogues are deemed irreplaceable, and there may be some doubt involved as to the quality and completeness of the material offered via terminal. Some commercial reference systems in USA are indeed still deficient in completeness and in cross-references. Yet the next generation of scientists - having had the computers already in the cradles - will have to depend on the terminal along with the microfiche reader, and these tools will enable them to work with better efficiency and yet with the same thoroughness, in comparison with the potential of earlier gnerations. The current literature on data retrieval is partly management-oriented, full of codes, acronyms and other documentese language, and not easy to read. However, we have user-oriented communications from some data centers, and retrieval references could gradually emerge as vital parts of any advanced textbook, much as bibliographies for recommended reading have long been already.

When preparation of this conference began a year and a half ago, suggestions for the agenda piled up so rapidly that we had to make a selection from the shopping list, lest we would have had to impose upon our hosts for another week. One of the items dropped from the agenda was "Financial Trends", on which - beyond a certainly very stringent awareness - not much can be done by way of coordination. I believe that something should be noted, and considered on the long run: It has been estimated that, even in highly industrialised countries, a few tenths of 1% of public Research and Development funds at best go into data processing and retrieval. At this level we cannot expect to do justice to the material to be processed, and the wealth of information is answered by a poverty of attention. It seems that 2% or 3% of R/D expenditures would be a more adequate figure, and still 10 times less than what science management requires. Compare this with the gains to be achieved if at least part of the time and costs for inadvertent and redundant duplication of data could be avoided through expedient access to pre-existing material. Needless to say, a larger retrieval clientele will also reduce the unit costs of search.

As to which structure the combination of databases will take, some thoughts offered by G.Wilkins at the close of the previous conference can be drawn upon. The data centers at the heart of the system are primarily charged with the responsibility for data of the publishable kind, that is, published at least under the standards of golden times before the invention of page charges. In addition there will be datafiles at research institutions, not present in the centers except by reference. These will have the least access demand, yet they will contain the occasionally needed backup information. It would usually still be uneconomical, for instance, to put photographic archives on tape. (This situation is comparable to the difference between abstracts, taped and disseminated by abstracting services, and full-length papers which require a subscription or a copying service. Some sciences outside Astronomy even got into the unfortunate position that full papers cannot be printed any more but have to be retrieved from a depository.) Thirdly, we have the derivative, selective, or critically evaluated data from the hands of the specialists of respective subject areas, a kind of synoptic literature. These files will receive most access requests, and from a most diversified and nonspecialist clientele. They should become fully banked, probably also accompanied by explanations of the compilation or by caveats of use. It is at this point that advice from designated specialists, Commissions, and Working Groups must be relied on most heavily, in the modification of accessible data, and also in the termination of access, should for instance a classification scheme or a set of standard objects be deleted in order not to cause future confusion. Where this cooperative effort has still been deficient (perhaps owing to inertia or to the desire to stay out of conflicts), we may proceed on the optimistic assumption that, as the importance of data services grows, so will the interest of the scientists to have the most reliable, complete, and error-free data material on record.

Newton's First Law states that everything in the world is determined by inertia, unless an effort toward change is made. Of course this applies only to the world of physical science and not to society. Yet we are far enough into the computer age that society does realise the advantages of modern information chanelling. The problems we are to negotiate are of profound influence on the structure of scientific communication. They relate to the spectrum of different customer demands, also to interscientific and international compatibility, standardisation, efficiency of management and guidelines, and far-sighted planning. Since problems, however, are not getting solved by reiterating their complexity, let me conclude, again appreciating your attendance here, and let us get to business.

A. EXISTING DATA CENTERS

THE STELLAR DATA CENTER

Jaschek C.
Centre de Données Stellaires - Strasbourg (France)

ABSTRACT

The paper presents an overview of the problems astronomical data centers are facing today.

When one organizes a colloquium, one can do some fairly outrageous things, like putting ones own contribution first. However as an excuse for such a procedure, I shall say that I want really not to discuss that much what the C.D.S. itself does, but what problems we have been confronted with, specially those which came up since the last Strasbourg colloquium, or which got more complicated to handle.

Until very recently, Data Centers were mostly distributors of magnetic tapes. This is a nice way of sending out information, if the user and the provider live in the same country and use compatible computers. If they live in different countries usually there exist mailing and custom problems. Tapes from Paris to Madrid have been found to have velocities of 20 km/day which is less than stage coach velocities of the XIX century. However recently direct access to data bases became feasible, in real time, like for instance through EURONET in Western Europe. One can say confidently that within a few years all major observatories shall be linked to some data center - regional or national - so that many operations which seem unrealistic today will be regarded as routine operations in a near future. For instance checking from the observers dome if the star one sees in the finder as double is really double or if one got the wrong star, or if the other star brighter than 9^m near the object one is observing is a nova or a known object. Similarly one will be able to ask from ones desk for all the known data about the star one wants to study, and for the list of all papers mentioning the object. Such a facility is already available at Strasbourg and can be seen in the adjoining room. This is the facility which was already provided to European astronomers working on I.U.E. at Madrid, and so far the reactions have been very positive - each space observer saves a lot of

C. Jaschek and W. Heintz (eds.), Automated Data Retrieval in Astronomy, 3–7.
Copyright © 1982 by D. Reidel Publishing Company.

time knowing which of the parameters of the object are already known from ground based astronomy. It is curious that several ground based observatories which have been offered the same facility as E.S.A., have hesitated to use it. Why ? Because they think these data should be obtained and paid by the observers before they use the telescope. Let us hope that in the next decade we can convince our colleagues of the usefulness of the facility !

It is sure that the direct interrogation of data bases in real time from any observatory constitutes an important advance in data accessibility and I am sure it will enhance the importance of the work done at the data centers. This is certainly an important point for all of us, since we know that up to now data collecting activities have not been exceedingly popular projects in many countries. There exists a widespread feeling that it is a somewhat second order activity. Although rightly no one could say that we are doing "frontier research" (as the people with the largest telescopes) it is important to remark that in natural sciences like astronomy, any single astronomer cannot afford to start from scratch - he has to rely to a very large extent upon the work of his predecessors or his colleagus. Therefore data center activities, which represent the collective memory should be regarded as being work of the same level of importance as any other scientific project.

One thing which makes data center activities less attractive to many people is the fact that it is largely anonymous. Whereas the author of a catalog is usually well remembered, people at data centers are usually oftener forgotten than remembered, and the ready accessibility to data via networks will contribute even a little bit more to this. If seems to me that one should insist explicitely on the recognition of the services provided. If it is common use to thank the directors of the observatory where one made his observations, or his computations, I do not see why one should not ask for a similar recognition if data centers or bases have been used extensively.

One can ask next as to what shall happen at the moment every one can interrogate data banks from his working place. Shall we insist on delivering only standard data of a given type, or should we condense or process the data further ? Up to now generally we have been content to store data and to disseminate them, but sooner or later the problem of condensing data will come up, specially if satellites provide us with many more data. We hope to learn more on this point on Thursday from the Director of the largest existing facility.

Another instance where data processing could be very useful is for surveys like Hα or 21 cm - here the possibility to generate isophotes would certainly enhance the value of the data. Now this question opens others, because if we do that, we will be asked to provide superposable displays of different kinds and so on and so forth. Thus it seems logic to start discussing where one should put the limit. This is closely related to the libraries of programs which are available at some computer centers. We discussed once the idea of having a library of astronomical programs, but our scientific council pointed out that the programs have to be either very simple (for the general user) or to be rather complicated (for the specialist). Since the specialist will very probably prefer to do his own program, the complicated programs are not very useful and therefore the whole idea of the library of programs collapsed. By now it seems to me that we should not mix our data base activities with too much additional work, but I would be happy if the subject be discussed at the meeting.

Let us leave now this area and shift our attention on what we have achieved so far. From the lists published in the reports of the data centers we know that there are about three hundred catalogs available on magnetic tape. It is important to recall that this is NOT the product of one center or even one country, quite to the contrary, these catalogs come from at least a dozen countries. I think that we should be proud of this international cooperation, reached through a number of bilateral agreements, which greatly facilitate the exchange of data. At the present time there are agreements relating the data centers in U.S., U.S.S.R., Japan, G.D.R. and we plan to extend this further, to arrive at the point where all astronomers from all countries can have access to ALL existing data. This was a dream, but thanks to your cooperation it is about becoming true.

Nevertheless, we should not be too self content, because first we have not yet on tape all what we need. To quote but a few examples of what is missing : the complete Durchmusterungen, the Astrographic Catalogs Luytens various surveys. Second, because not all countries have data centers, and use very little the existing facilities and thirdly, because we have not been able so far to update our catalogs and files with the necessary speed. Let me provide some examples. I start with a case familiar to me, the "The Bibliographic Star Index" : we are running behind, since we have only completed the literature to 1979, o. Well, we hope to do better in the next future, but that is something everybody hopes to do Photometry also is a rapidly moving field, and infrared catalogues are some years behind, as also u v b y photometric catalogues. In the latter case the annual growth is of the order of 20 %, counting only the number of new stars added. Similarly in spectral classification we have not yet on tape the excellent work done by our Soviet colleagues ; also here the number of stars classified is very large. In astrometry we have still nothing better than S.A.O. for positions and proper motions over the whole sky, despite the fact that the material for such an enlarged S.A.O. exists - even on magnetic tape. And let us not forget some astrophysical parameters like magnetic fields, whose measurements were never collected in a catalog.

I think that the data centers should play a more active role in astronomy, in the sense that they should encourage people to do work which is needed. This implies contacting the best specialists in the field and ask them to do the job, be it a catalog, a selection or a bibliography. I insist strongly, and you should excuse me, on the principle that a good catalog can only be made by a good specialist. It is quite useless (but happens unfortunately) that colleagues who never used a given technique compose a catalogue using this technique. This I feel is a bad practice, since surely he will not be aware of the many pitfalls on the road, and produce something of inferior quality. So in my idea, data centers should attempt to be ahead of developments, trying to develop new catalogs and/or bibliographies. I know that this involves a great deal of sometimes frustrating work, but I think that those colleagues who work at data centers have a much better feeling for what is missing than the average astronomer. Let me also insist that data centers should try also to act as diffusers of the specialities being developed in their country. This means simply that if in the country X there is a strong development of the speciality Y, then the local data center should feel obliged to see that the results of this development reaches the mainstream of astronomical information. Curiously this simple idea needs much implementation in many countries, specially in those in which astronomy is just beginning to be developed.

If we consider next what happens in data activities in our field, I think that we have been organizing successfully the stellar area mainly. This is certainly a reasonable approach, since some problems are much less complicated for stars than for non-stellar objects. A first obvious problem is the existence of instrumentally defined nomenclatures, for instance in the case of an HII region with dimensions defined very much by the resolution of the receiver. If one changes the receiver, the picture alters in a significant way and the problem arises if one is dealing with exactly the same object. Multiple names attached to the same entity, or to different parts of the same objects occur frequently and there is no clear cut resolution to the problem. Probably all these objects need a detailed cartography together with the nomenclature and the only way to produce this is through images on microfiches. We hope to hear more on this from those colleagus involved in non-stellar data centers. There is one important point however which I hope shall not forgotten in the discussion, namely that even more than for stars we need detailed bibliographics of each object. Whereas in stars many parameters can be catalogued because they can be summarized in a compact way, this is not so in the non-stellar area. Just as an example think of the "diameter of a star" as opposed to the "diamater of a galaxy". The only reasonable solution seems to be a complete bibliography of the galaxy one is interested in, so that one can get the papers which contain that part of the information which cannot easily be summarized.

One should also think of the creation of a data bank for the solar system. I am sure that a bibliographic file on each comet or asteroid would be greatly appreciated. But is somebody doing such a thing ?

A last point I want to consider is the question of how to publish the catalogs : i.e. to what could be called "editorial policies". It is true that we are primarily interested in computer readable data, but please do not forget that first of all we are astronomers and as such working for the future generations, which implies that we have to preserve our observations. How long can a tape be preserved, if stocked properly and re-read say, once a year ? I have found different opinious but nobody said that they can be stored without alterations more than ten years, with the present technology. Now this is nothing in face of the time over which an astronomical observation is used. Probably we shall hear more on this subject in the bibliographic section, but let me anticipate that we cannot liberate us yet from the need of a permanent support, like paper, film or plastic. You can answer that scientific magazines are just there to print catalogs. But if you consider the page charges which increase and the typography which decreases in some journals with time, you feel that catalogs cannot be any more printed the traditional way, if you want to have them updated frequently. This is a "must" in certain fields, like photometry, where one gets an increase in the number of objects by 10 - 20 % per year. So if you do not want to have a large number of data "lost" for pratical purposes because they are spread out over two hundred publications, you must find a "quick" solution and that is why I believe that the microfiches are to stay with us, despite the inconvenience of the need of a microfiche reader. To underline this, just consider the case of any printed catalog. The traditional way of a type-set catalog produces an inevitable delay of two years at least after the catalog is compiled, so that by the time it is distributed, it is already outdated. We need definitely to remedy this and I hope we shall hear more on this topic in the session on "Editorial policies".

I shall stop here ; I have listed many open questions and I hope that this meeting shall help to clarify them or to produce answers.

THE ASTRONOMICAL DATA BASE AND RETRIEVAL SYSTEM AT NASA

Jaylee M. Mead
Laboratory for Astronomy and Solar Physics, NASA/GSFC

Theresa A. Nagy and Robert S. Hill
Systems and Applied Sciences Corporation

Wayne H. Warren Jr.
National Space Science Data Center, NASA/GSFC

ABSTRACT

More than 250 machine-readable catalogs of stars and extended celestial objects are now available at the NASA/Goddard Space Flight Center (GSFC) as the result of over a decade of catalog acquisition, verification and documentation. Retrieval programs are described which permit the user to obtain from a remote terminal bibliographical listings for stars; to find all celestial objects from a given list that are within a defined angular separation from each object in another list; to plot celestial objects on overlays for sky survey plate areas; and to search selected catalogs for objects by criteria of position, identification number, magnitude or spectral type.

INTRODUCTION

There are two major objectives for the development of the astronomical data base and retrieval system at the NASA/Goddard Space Flight Center (GSFC): (1) the acquisition, verification and documentation of all available machine-readable catalogs of stars and extended celestial objects; and (2) the creation of retrieval techniques which will enable a user to utilize these data easily and efficiently from a remote terminal. More than 250 computerized astronomical catalogs are now available at GSFC. Periodically reports are published in the Astronomical Data Center Bulletin giving the status of verification, documentation and hence availability, for each catalog in the data base (Nagy et al. 1981b). Because the output of any retrieval system is only as good as the data base from which it obtains its information, much of our effort continues to be devoted to the improvement of the individual catalogs, as described by Warren et al. (1982).

C. Jaschek and W. Heintz (eds.), Automated Data Retrieval in Astronomy, 9–13.
Copyright © 1982 by D. Reidel Publishing Company.

Three groups at Goddard are primarily responsible for the data handled by the Astronomical Data Center (ADC): the National Space Science Data Center (NSSDC); the Infrared Astrophysics Section of the Laboratory for Extraterrestrial Physics; and the Laboratory for Astronomy and Solar Physics, which also includes the newly established International Ultraviolet Explorer (IUE) Regional Data Analysis Facility.

The NSSDC, whose data responsibilities are reviewed by J. Vette (1981), serves as a distribution center for astronomical catalogs. The high-energy astrophysics data sets are described by M. Locke (1981). W. Warren Jr. is responsible for catalog data at longer wavelengths as well as archival and dissemination of observations from the IUE satellite (Warren and Alderman 1982).

The Infrared (IR) Astrophysics Section is committed to producing an IR (1-1000 μm) data base, bibliography and cross index of IR names as described by Schmitz et al. (1982).

The Laboratory for Astronomy and Solar Physics, with which T. Nagy, R. Hill and J. Mead are associated, does much of the verification and documentation of catalogs before they are placed with the NSSDC for distribution. The primary interest of this group is the creation and implementation of computerized retrieval systems described below.

The development of the data base and retrieval tools in a location where there is daily interaction with other astronomers promotes the familiarity of data personnel with the needs of both observers and theoreticians. It also encourages rapid evaluation of data products while providing many consultants on specialized astronomical and data handling topics.

ASTRONOMICAL DATA BASE

Part of the success in acquiring the large number of catalogs in the ADC data base is due to the Cooperative Agreement between NASA and the Centre de Données Stellaires (CDS), which was worked out as a by-product of the IAU Colloquium 35, held at Strasbourg in 1976. This agreement has also yielded the beneficial results of exchanges of error lists, personnel and information about future projects (Mead 1980; Mead et al. 1981).

Not all machine-readable catalogs received by the ADC are ready to be used. The first step in verification is to try to read the catalog as specified in the documentation accompanying it. This procedure quickly exposes most problems--if not in the machine-readable data, then sometimes in the format descriptions. Suggestions for formatting, checking and documenting machine-readable astronomical catalogs are given in detail by Hill (1981).

RETRIEVAL OF DATA

One of the most important aspects of a data base is its capability for efficient retrieval of the data. Six of the most frequently used astronomical data retrieval tools now available at the ADC are described below:

(1) The Interactive Computer Reference Search of the Astronomical Literature 1950-1976 (Nagy et al. 1981a) uses the data and reference files of the 1976 version of the Bibliographical Star Index (BSI) (Ochsenbein and Spite 1977). These data are stored on direct access devices on the GSFC IBM S/360-91 computer. The BSI is a machine-readable data file of stellar identifications and references covering twelve journals from 1950-1972 and more than 30 since then. Updated versions of this compilation are released periodically by the CDS. The machine-readable version available at Goddard contains 9494 references to the astronomical literature for 69,348 stars from 1950 partway through 1976; a later version of the BSI will be utilized as soon as the tape is received from the CDS.

The interactive BSI search program accesses the references by means of a star identifier (Henry Draper (HD) or Durchmusterung (DM) catalog number, or variable star name). The BSI references for the stars having these identifiers are then displayed at the user's computer terminal in a form resembling that of a standard bibliography. One advantage of this method of accessing the BSI lies in the conciseness of the output: only one session at a data terminal is needed to produce a combined, chronologically-ordered listing of references from all the years covered in the BSI. Another advantage lies in the ease of use: one only has to type in the stellar identifications according to instructions displayed on the screen by the program.

A limited amount of computer time has been set aside for a pilot program to utilize the BSI search program. All astronomers have been urged to avail themselves of this opportunity to access the data set from their home institutions by telephone link. Many favorable reports have been received from users.

(2) The MATCH Program (Hill and Nagy 1981) finds all objects in a list within a user-defined angular distance of the positions of a set of target objects. This technique is particularly useful when one wishes to make identifications of objects observed at similar wavelengths or at very different ones, as often occurs when observing in a survey mode. In addition, this program serves as a useful check on catalog identifications which have been made through manual comparisons, and it provides guide star candidates for observing programs where the objects of interest are very faint and/or have poor positions. Catalogs which have been used in MATCH runs include the following combinations: the Two-Micron Sky Survey vs. the SAO Catalog; the General Catalogue of Variable Stars vs. the Equatorial IR Catalogue (EIC 2); and the Dearborn Observatory Catalogue of Faint Red Stars vs. the EIC 2.

(3) The Job Control Language (JCL) Copy Program (Hill and Nagy 1980) facilitates the duplicating of star catalog tapes by supplying all the necessary JCL parameters, such as block size, density, record length, etc., requiring only a four-digit numerical code for the catalog to be entered. Not only is this program time-saving in setting up tape copies, but it also reduces failures in the duplicating of tapes by automatically providing the information needed to copy the tape.

(4) Two search and retrieval programs have been developed (Warren and Sheridan 1981) for the Catalogue of Stellar Identifications (CSI) (Ochsenbein et al. 1977), which contains cross references for star numbers from some 30 different catalogs. Following the preparation of a version of the CSI sorted by increasing right ascension, a program was designed which accepts equatorial coordinate limits and lists all CSI stars located within the specified region. This search mode is useful for selecting possible optical identifications for a list of sources which are detected at other wavelengths or which are variable.

A second search program accepts DM and HD catalog numbers and lists all CSI stars with the specified input identifiers belonging to any of the various catalogs connected to the CSI. Individual data sets are created and stored for each catalog desired, so that each one can be read by a subsequent program to retrieve data from the corresponding source catalog. Ideally, this retrieval system will develop into a tool by which data from any number of source catalogs can be selected in a single run through a link to the CSI.

(5) Five catalogs (SAO Star Catalog, Revised New General Catalogue of Non-Stellar Astronomical Objects, Reference Catalogue of Bright Galaxies, Two-Micron Sky Survey and CSI) have been sorted by Palomar (and ESO/SRC) Sky Survey plate areas, as described by Mead and Nagy (1977). For a given set of coordinates the computer provides all the plate numbers on which this position can be found. These plate areas can be immediately accessed, and listings and plots of any or all of the objects from the five catalogs generated.

(6) Software has been developed to retrieve the full data entry for HD stars from any of the eleven catalogs included in the Goddard Cross Index in a single computer run (Mead and Nagy 1977). Another cross index which has proved to be particularly useful is the HD-DM-SAO Cross Index (Nagy and Mead 1978).

CONCLUSION

The ADC is continuing to provide these and other data retrieval routines to GSFC astronomers and to the astronomical community on a request basis. With an upgraded Goddard computer available later this year, we expect to have more data sets on line, such as the CSI, in order to make interactive retrieval systems readily available to anyone who wishes to access data directly by dialing up the GSFC computer.

REFERENCES

Hill, R.S.: 1981, Suggestions for Formatting, Checking and Documenting Machine-Readable Astronomical Catalogues, ADC Bull. 1, pp. 69-77.

Hill, R.S., and Nagy, T.A.: 1980, User's Guide for Program ACPYJCL, EAC-T-5294-0030.

Hill, R.S., and Nagy, T.A.: 1981, Documentation and User's Guide for the MATCH Program, EAC-T-5294-0031.

Locke, M.C.: 1982, NSSDC and WDCA for Rockets and Satellites: Their Role in X-Ray Data, IAU Coll. 64, Automated Data Retrieval in Astronomy, this volume, p. 45.

Mead, J.: 1980, On-Going and Future Astronomical Data Base Activities at the Goddard Space Flight Center - February 1980, Bull. Inform. CDS No. 19, pp. 57-59.

Mead, J.M. and Nagy, T.A.: 1977, Retrieval Techniques and Graphics Displays Using a Computerized Stellar Data Base, IAU Coll. 35, Compilation, Critical Evaluation, and Distribution of Stellar Data, pp. 161-166.

Mead, J.M., Nagy, T.A., and Warren, W.H. Jr.: 1981, On-Going and Completed Astronomical Data Base Projects at the Goddard Space Flight Center, November 1980, Bull. Inform. CDS, No. 20, p. 78.

Nagy, T.A., and Mead, J.M.: 1978, HD-DM-SAO Cross Index, NASA TM-79564.

Nagy, T.A., Hill, R.S., and Mead, J.M.: 1981a, Availability of an Interactive Computer Reference Search of the Astronomical Literature 1950-1976, ADC Bull. 1, pp. 62-68.

Nagy, T.A., Mead, J.M., and Warren, W.H. Jr.: 1981b, Status Report on Machine-Readable Astronomical Catalogues, ADC Bull. 1, pp.146-169.

Ochsenbein, F., and Spite, F.: 1977, Main Features of the Stellar Bibliographic File, IAU Coll. 35, Compilation, Critical Evaluation, and Distribution of Stellar Data, pp. 175-177.

Ochsenbein, F., Egret, D., and Bischoff, M.: 1977, The Catalogue of Stellar Identifications, ibid., pp. 31-36.

Schmitz, M., Gezari, D.Y., and Mead, J.M.: 1982, Infrared Astronomical Data Base and Catalog of Infrared Observations, IAU Coll. 64, Automated Data Retrieval in Astronomy, this volume, p. 33.

Vette, J.: 1982, Data in Space Astronomy, ibid., this volume, p. 243.

Warren, W.H. Jr., Mead, J.M., and Nagy, T.A.: 1982, Responsibilities and Practical Limitations in the Operation of an Astronomical Data Center, ibid., this volume, p. 131.

Warren, W.H. Jr., and Alderman, D.F.: 1982, Archiving and Retrieval of Image Data for the International Ultraviolet Explorer (IUE) Mission, ibid., this volume, p. 115.

Warren, W.H. Jr., and Sheridan, W.T.: 1981, Search Capabilities for the Catalogue of Stellar Identifications (CSI) 1979 Version, ADC Bull. 1, pp. 114-120.

THE SOVIET CENTER OF ASTRONOMICAL DATA

Dluzhnevskaya O.B.
Astronomical Council - U.S.S.R. Academy of Sciences
Moscow

On the basis of the current French-Soviet cooperation in science and technology, the Astronomical Council of the U.S.S.R. Academy of Sciences and the Strasbourg Center signed in 1977 an agreement on setting up the Soviet Center of Astronomical Data as its filial branch. The Soviet Center was created on the basis of a computation center at the Zvenigorod station of the Astronomical Council of the U.S.S.R. Academy of Sciences, which had already had considerable experience of working with stellar catalogues. In 1979 the Center was equipped with a EC-1033 computer.

In 1978-1979 the Soviet Center of Astronomical Data (C.A.D.) received from Strasbourg 96 of the most important catalogues. By September 1981 the list of catalogues available at the Soviet Center has reached 140 catalogues.

The Soviet Center has the following aims :

1. to obtain from Strasbourg, copy, keep and distribute in the U.S.S.R. the stellar catalogues and other astronomical data recorded on magnetic tape ;

2. the preparation, tape-recording, technical control and exchange with Strasbourg of the astronomical catalogues compiled by Soviet astronomers ;

3. to undertake astronomical investigations and research work based on data available.

As has been mentioned, the main tasks are those of accumulating, keeping and computer-processing of astronomical data files with the purpose of their subsequent distribution in the U.S.S.R., and centralizing the data exchange with Strasbourg. At present 140 catalogues are included into the files. All the data are recorded on EC computer magnetic tapes (the volume of information on one tape may reach 20-22 megabytes).

C. Jaschek and W. Heintz (eds.), Automated Data Retrieval in Astronomy, 15–20.
Copyright © 1982 by D. Reidel Publishing Company.

The work with catalogues is carried out on an EC-1033 computer at the Astronomical Council of the U. S. S.R. Academy of Sciences. The computer speed is 180000 operations per second, the working storage is 512 kylobytes. The external memory includes magnetic discs of 7.25 megabytes and magnetic tapes with a record density of 32 bytes per mm (800 bit per inch). The computational process is controlled by an operational system OC Ec.

All the available catalogues are systematized in agreement with the classification of the Strasbourg Center. Due to limited possibilities in manpower and technological means the C.A.D. cannot cover all branches of astronomy. We have in view first of all the branches covered by the Strasbourg Center and even in them we have so far given preference to the fields close to the research interests of the Astronomical Council namely photometry and spectroscopy of stars and stellar systems, data on non-stellar objects (radio sources, nebulae, etc...). In the field of astrometry the C.A.D. needs several large fundamental catalogues, in particular, for solving problems in satellite geodesy. The main work in the critical analysis and distribution of astrometric data is carried out by the Center of Astrometric Data at the Main Astronomical Observatory (Pulkovo) of the U. S. S.R. Academy of Sciences. We have also included, or are going to include, in our files various tables of astrophysical interest : tables of oscillator strength, tables of stellar opacities, etc... Besides, the files will include catalogues of photographic and laser-ranging observational data on satellites which are being compiled now. So far we do not have any data on Solar system objects. A small part of this information may also be included in our files later. Recently we have received a magnetic tape with data on solar activity from the World Center of Geophysical Data at the Joint Geophysical Committee of the U. S. S.R. Academy of Sciences.

One of the most important aspects in the activity of the Center is the service which is being set up for correcting errors found in published and machine-readable versions of catalogues.

Specialized software was developed at the C.A.D. It includes several isolated programs and the "SPARTAK" system which, with the help of a set of simple instructions, makes it possible, for example, to prepare various catalogues samples and to keep a record of entering inquiries.

Further development of the software is now aimed at a complex utilization of the available data files. A catalogue of star identification (of the Strasbourg type) will serve as a basis of this astronomical data base, which will solve the problem of cross references between different catalogues.

One of C.A.D.'s main tasks is the distribution of astronomical data presented on computer media in the U. S. S.R. Information on the services offered by the Center and the available opportunities of work with data files is regularly circulated to all observatories. At the present the

CAD upon request copies data or prepares various samples, provides assistance and helps in a unified preparation of Soviet astronomical catalogues. In 1980 the CAD granted 17 requests of different organizations for copying and processing of 68 catalogues on magnetic tapes. As a rule a request is granted within a month.

Another important task of the CAD is the centralized preparation of Soviet astronomical catalogues on magnetic tapes and subsequent exchange of these data with the Strasbourg Center. Advice is given to observatories which can tape-record their catalogues themselves.

By now, with the help of the CAD staff, the following Soviet astronomical catalogues in the machine-readable form have been prepared and sent to the Strasbourg Center of Stellar Data :

1. A.E. Piskunov, "Catalogue of Stellar Masses and Ages for 68 Open Clusters", where evolutionary masses and ages of 7000 stars are given determined from stellar positions on the H-R theoretical diagram.

2. Z. Kraitcheva, E.I. Popova, A.V. Tutukov, L.R. Yungelson, "Catalogue of physical parameters of spectroscopic binaries", in which besides the data given in Batten's Catalogue (6th edition) and its extensions, the following stellar characteristics are determined : masses and mass ratios of components, semiaxes of orbits, orbital angular momentum, specific orbital maomentum.

3. R.A. Bartaya, "Catalogue of spectral types and luminosity classes of 10396 stars in Kapteyns areas Nos. 2-43".

4. Z. Alksne, A. Alksnis, "Catalogue of cool carbon stars discovered at the Radio Astrophysical Observatory of the Latvian Academy of Sciences" (219 stars).

5. M. Muminov, "Proper motions and UBV-photometry of stars in the field of the clusters h and χ Persei".

6. V.N. Frolov, "Photometry and proper motions of stars of four open clusters in Cassiopeia".

Several other Soviet catalogues are under preparation for being recorded on the EC computer magnetic tape :

1. M.A. Svechnikov, "Catalogue of orbital elements, masses and luminosities of close binary stars" (about 200 systems).

2. N.G. Kogoshvili, "Morphological catalogue of galaxies".

3. P.N. Kholopov et al., "New catalogue of suspected variable stars". This catalogue contains data on 15000 stars and is already tape-recorded with the help of an BESM-4M computer. Its rewriting to fit the international standards is being completed.

4. P.N. Kholopov et al., "General catalogue of variable stars" (4th edition) where data on 40000 stars will be collected.

According to the Agreement between the Astronomical Council of the U.S.S.R. Academy of Sciences, the Budapest University (Hungary), and the Astronomical Institute of the Czechoslovakian Academy of Sciences, the Soviet Center is tape-recording the Catalogue of clusters and associations by B. Balasz, J. Ruprecht and V. Vanysek. The part of the catalogue containing data on associations is already tape-recorded and will be sent to Strasbourg in the near future.

Recently the CAD has prepared a list of existing Soviet catalogues in astrophysics and after a critical analysis some of these catalogues will be tape-recorded and also sent to Strasbourg.

There exists a large number of astronomical problems to solve for which it is necessary to use catalogue data. The use of computers makes it possible to improve considerably the efficiency of this work, since for example it guarantees against accidental errors made during data processing, allows and easy repeatable processing of the material (which is rather important for large catalogues containing several tens or even hundreds of thousands of stars), allows to include into the procedure of data analysis complex reduction algorithms which cannot be used in hand processing, and finally it is promising for mass utilization of combined data obtained merging several different catalogues.

Investigations have been carried out at the Astronomical Council using stellar catalogues in machine-readable form, including catalogues prepared by the CAD (M0001, M0002). Thus, paper /1/ devoted to the study of distributions of stars in open clusters over mass m and age t is based on the catalogue of m and t values for about 8 000 stars in 68 clusters (M001). Since stars of open clusters have common origin and evolutionary history, the study of mass and age functions in clusters can provide valuable information on peculiarities of the star formation process in these groups.

Using the catalogue of physical parameters for spectroscopic binaries (M002), Z.T. Kraitcheva, E.I. Popova, A.V. Tutukov and L.R. Yungelson have studied the distribution of these stars over component masses, mass ratios, periods, orbital momenta, and also studied correlations between the above quantities /2,3/.

The same authors have studied the sample of eclipsing binary stars with $V \leq 10^{m}.5$ using the catalogue by Wood et al. /4/ and shown that for these stars basically the same laws are observed as for spectroscopic systems /5/.

Papers /6/ and /7/ are devoted to the study of mass function peculiarities for different objects of the disk galactic population and belong also to the studies mentioned above.

Paper /6/ discusses the question if the initial mass function of population I type stars is monotonic. Analysis of a number of catalogues received from the Strasbourg Center (Michigan Catalogue of Spectral Classification S 3031, Philip and Egret Catalogue based on uvbyβ-photometric data S 5014, altogether 20000 stars) and also of several other catalogues, has shown that there is a number of independent evidences that the initial mass function of stars is nonmonotonic.

Paper /7/ is devoted to the study of the distribution of B-stars over mass in the association Orion OBI. For this sample, Strömgren photometric data of the Warren and Hesser Catalogue S 2036 have been used for about 500 stars.

Some aspects of the chemical evolution of the galactic disc have been studied from the data of catalogues S 5014 and S 3032 in papers /8/, /9/, and /10/. In paper /8/ an integral distribution is obtained for the dwarfs of the S 3032 catalogue over metallicity indices (Fe/H) determined by the spectral method.

The study of various catalogues is being carried out at the Astronomical Council. In particular, the present mass distribution is investigated for population I stars from the solar vicinity over a wide range of masses ($M_{\odot}$~0.3 - 30 $M_{\odot}$). To do so, we use data from several spectral, photometric and combined catalogues (S 3011, S 5014, S 5004, S 5001, S 5051, etc...) containing altogether up to hundred thousands of stars.

In the immediate future the Center will continue and deepen ist activity in all directions.

The establishment of the Soviet Center of Astronomical Data is an important event for Soviet astronomers whom it renders systematic assistance in obtaining and partial processing machine-readable astronomical data. Scientists of socialist countries are also interested in its activity and we hope that in the near future many interesting catalogues will be added to the data bank. The main data bank has been kindly placed at the disposal of the Soviet Center by the Strasbourg C.D.S. for which we are very grateful.

REFERENCES

1. A.E. Piskunov, Nauchn. Inf. Astron. Council of USSR Ac. Sci. 37, 47 (1976)

2. Z.T. Kraitcheva, E.I. Popova, A.V. Tutukov, 1.R. Yungelson, Astron. Zn., 55, 1176, (1978)

3. Z.T. Kraitcheva, E.I. Popova, A.V. Tutukov, L.R. Yungelson, Astron. Zh., 56, 520, (1979)

4. F.B. Wood, J.P. Oliver, D.R. Florkowski, R.H. Koch, Publ. Dept.

Astron. Univ. Florida, v. 1, (1980)

5. Z.T. Kraitcheva, E.I. Popova, A.V. Tutukov, L.R. Yungelson, Pisma Astron. Zh. (1981) (In press)

6. A.E. Piskunov, A.V. Tutukov, L.R. Yungelson, Pisma Astron. Zh., 5, 81 (1979)

7. N.M. Ageicheva, graduation paper, Ural State Univ., Sverdlosk (1981)

8. S.V. Vereshchagin, A.E. Piskunov, Astron. Circ. N° 1033 (1979)

9. S.V. Vereshchagin, Proceedings of The Meeting on Chemical Composition and Evolution of Stars, Baku, (1980) (in press)

MANAGEMENT OF ASTRONOMICAL DATA AT KANAZAWA DATA CENTER

Yoichi Terashita
Kanazawa Institute of Technology, Kanazawa 921, Japan

ABSTRACT. The methods of controlling and disseminating astronomical data at the Kanazawa Data Center (operated under the Computer Center of the Kanazawa Institute of Technology) are described. Discussions include the coded data descriptions used in conjunction with a file processing language. Problems concerning remote data access are also discussed, and the future plan presented.

1. INTRODUCTION

Since 1978, the Computer Center of the Kanazawa Institute of Technology (KIT) has been engaged in the task of disseminating machine-readable astronomical data to the astronomical community in Japan. The basis of the data service is a collection of data files (star catalogues) supplied by the Strasbourg Stellar Data Center (CDS), but locally produced data files are also being added. The current holdings at KIT is about 200 catalogs (about 300 data files if subfiles are counted). The main task so far has been to supply the user with duplicates (on magnetic tapes) of files, or with their printed versions. A scheme is being developed and experimented, such that the user can access and manupilate the data files through an on-line terminal.

These data files naturally differ in data items involved, in quantitative units, in recording format, and in descriptive symbols. While this type of diversity could be bearable when they are to be handled manually, it brings about a major difficulty in the case of computer processing. It was necessary, therefore, to develop a data management system specialized for astronomical data, by means of which one can handle data files with the least possible degree of complication.

In developing such a system, the emphasis has been in the application

C. Jaschek and W. Heintz (eds.), Automated Data Retrieval in Astronomy, 21–25.
Copyright © 1982 by D. Reidel Publishing Company.

of data management technologies that are currently available. We have been particulaly interested in those languages specialized for data retrieval and file processing. Some of the so-called data base management systems (DBMS) have also been examined and tested, but these are not discussed here.

2. ASTRONOMICAL DATA MANAGEMENT SYSTEM

Figure 1 shows the overall structure of the Astronomical Data Management System (ADMS) developed at the KIT Computer Center. Central to its design are three sets of system files that contain control information on the data files. The data directory (DD) contains the most basic pieces of infomation on the individual data files, such as the file number, file name, record length, number of records, and the storage infomation. The human-readable data description (HDD) contains the information on the contents of a specitic data file, that is, the position of each data item in the record, its description, unit, and other supplementary information. In addition, each data item is assigned a mnemonic code (field name), with which one can compose a file processing job, as described in the below. The HDD is prepared for each data file, and its direct print-out becomes a user reference manual for the file.

The coded data description (CDD) is obtained from the corresponding HDD through a conversion routine. A CDD defines, for each data item in a given file, the field name (the same as that assigned in the HDD), the location in the record, and the type (numeric or character string), in accordance with the syntax of the language that has been selected as the standard file processing language of the system. We can select any language as long as it is so designed that the data definition part of a program can be separately prepared. At present, a language called EASYTRIEVE is used in ADMS.

Functions of ADMS are summarized in Figure 1. When the user wishes to work on a data file(s), he feeds in the file number through an on-line terminal. The system then retrieves the relevant storage information for the file from DD (Step 1) and incorporates the corresponding CDD (Step 2). The user then types in EASYTRIEVE instructions, consulting the printed version of HDD and using the field names defined there (Step 3). Normally this amounts to only a few lines of input, since only procedural specifications are necessary. After the job is activated and processes the designated file (Step 4), the resulting output is returned to the terminal (Step 5). Figure 2 shows a sample session involving the Yale Catalogue of Bright Stars (file number C50021). Lines with lower-case letters are those fed in by the user. In this example, the job is to list the magnitude, position, and name for the six brightest stars with declination greater than +40 degrees.

Although it would certainly be desirable to keep all the data files

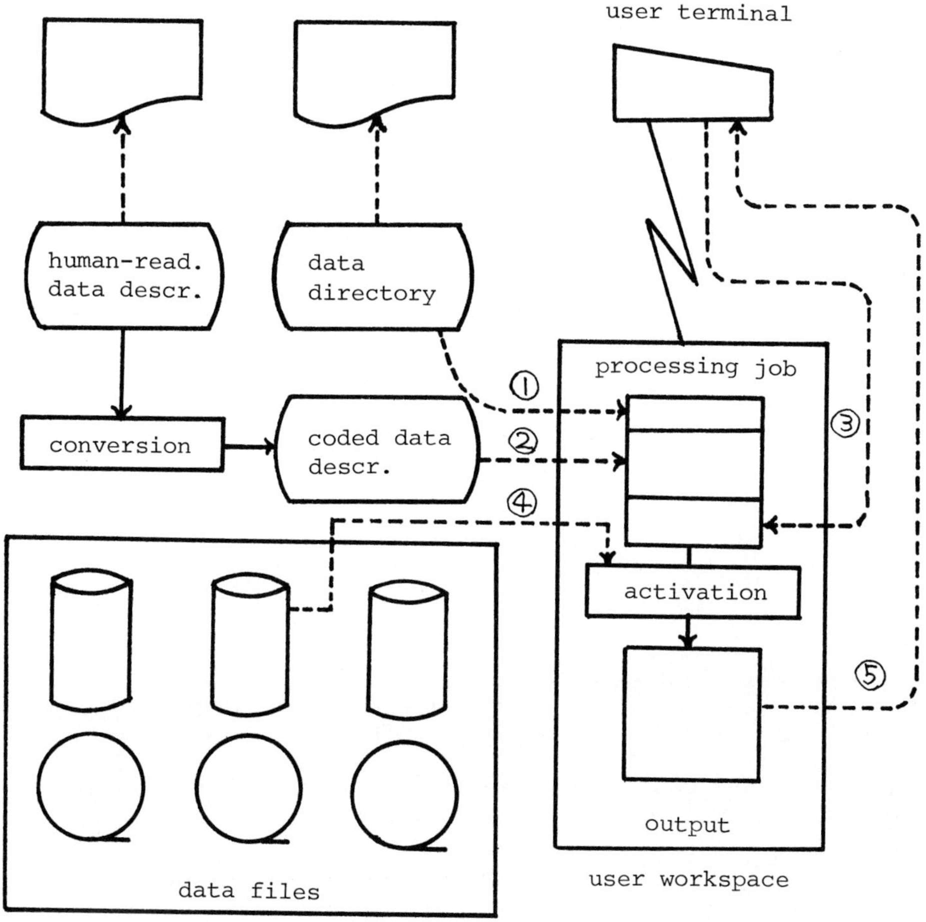

Figure 1. Structure of the Astronomical Data Management System.

on-line, only about half of them are currently resident on disks, for the reason of economy. Nevertheless, files on off-line tapes can still be accessed in the same fashion, though, in this case, assistance from the computer operator is required.

3. FILE PROCESSING LANGUAGES

While almost any computer language can be used to process high-volume data files such as star catalogs, the amount of programming effort for a given task varies depending on what language one should choose. Thus, it can safely be said that COBOL and PL/I are more useful than FORTRAN and ALGOL. However, the former languages are not necessarily the best. There are now available a certain number of languages specifically designed for file-processing applications. Unlike

```
.qastro c50021 example
*** DISK OR TAPE ? ('D' OR 'T')
.d
*** CATALOG C50021 READY FOR PROCESSING ***
INPUT:
.9999
.if del20deg gt 40
.sort mv
.list lj limit 6 seq alp20 del20 mv name
.
*** SUBMIT ?
.yes

.elist d

 7/01/81
  SEQ     ALP20     DEL20       MV        NAME
  1708     51641     46  0        9     1301  AUR
  7924    204126     45 16      126     5001  CYG
  4905    1254 2     55 57      176     7705  UMA
  1017     32420     49 51      179     3301  PER
  4301    11 344     61 45      179     5001  UMA
  5191    134732     49 19      186     8507  UMA
          6   RECORDS PRINTED
```

Figure 2. Sample session using ADMS.

general-purpose programming languages such as COBOL and PL/I, these languages offer simplified sets of instructions in such a way that one no longer needs to specify details of operations (the non-procedural feature). This feature is especially noticiable for the looping structure, sort operation, and output editing: that is, the operations always needed in file processing jobs.

EASYTRIEVE, which is currently used in ADMS, is one of such languages. Before this one, we used GIS, another (and one of the earliest) of them (see Ref.1). FOCUS is one of the more recently developed, which we are interested in but has not yet been tested. Statistical languages, such as SPSS and SAS, can also be used for file processing, but their direction is somewhat different and they usually lack in efficiency (we plan to connect SAS to ADMS anyway, since statistical processing is, as a matter of fact, highly demanded in astronomical works).

These file processing languages are not without their problems. Their main purpose has been for business applications. Thus, FORTRAN-like numeric types are not popular among them (standard data types are packed- and zoned-decimals). In business applications it is usually taken for granted that data files are well organized and stored in accordance with established standards, while this is not the case for astronomical (or scientific in general) data files. In the latter case, it is often necessary to perform operations involving two or more files concurrently, while such operations are not always easy with the languages mentioned here.

Although it is certainly advantageous to use such languages, our conviction has been that there should be file-processing languages that are more scientifically oriented than those now available. As an attemt, one such language is being developed at the KIT Computer Center.

4. DATA COMMUNICATIONS FACILITIES

The means of data dissemination at Kanazawa has been predominantly off-line, that is, magnetic tapes and printed outputs (produced under ADMS). On-line access to Kanazawa data files by way of public telephone lines has been experimented, but the results are not very encouraging. One reason is the quality of the line, which exhibits considerable variations depending on the geographical location of the caller. Another is the cost for such a connection (charged as a usual long-distance call).

The only solution to these problems seems to be to use a dedicated digital communications network. In Japan, the DDX system, a packet-switching digital network developed by the Nippon Telephone and Telegraph Corp. (NTT), is now available at certain areas of the country, and is expected to cover the Kanazawa area in the near future. An interface to the DDX-based N-1 Network (a nation-wide inter-university computer network) has been developed at the KIT Computer Center, and is being tested.

We are indebted to colleagues at CDS for their continuing assistance and encouragements.

REFERENCE

1) Terashita, Y. 1981, in 'Scientific Information Systems in Japan' (Ed. H. Inose), pp.164-169, North-Holland Publ. Co., Amsterdam.

ACTIVITIES OF THE POTSDAM DATA CENTER

Schilbach E.
Central Institute for Astrophysics of the Academy of Sciences
Potsdam - German Democratic Republic

In 1968 the Data Centre was established in the Central Institute for Astrophysics of the Academy of Sciences(CIAP) of the German Democratic Republic. Its task was essentially the analysis of astronomical observations and the elaboration of theoretical models for astronomical problems.

In 1974,a cooperation was started between our institution and the Stellar Data Centre at Strasbourg.

As a consequence, in the Data Centre a working group of three scientists was formed, whose task was to collect, store and process astronomical catalogues on magnetic tapes and to forward these data to astronomers in socialist countries. The CIAP has a computer ES 1040 with the following characteristics :

Mean Operating speed	:	380 000 op/sec.
Main storage	:	512 Kilobytes
Mass storage	:	magnetic tapes, 9 traks, 800 bpi recording techniques : NRZI magnetic discs volume : 28 Mb
Operating system	:	OS - ES
Programming languages	:	Algol, Fortran, PL/1, Assembler.

In cooperation with the CDS, the CIAP supplies data to Strasbourg for the "Bibliographical Catalogue of Variable Stars" (CDS Number 6017). At the moment about 257 000 bibliographical records have been prepared on magnetic tape and sent to the CDS. The catalogue should be ready by January 1982 and it will be supplemented each year.

This work is described in detail in CDS Bulletin 20, 105 (1981) in a paper by Dr. Wenzel.

The basic stock of catalogues in our Data Center comes from the CDS.

C. Jaschek and W. Heintz (eds.), Automated Data Retrieval in Astronomy, 27–28.
Copyright © 1982 by D. Reidel Publishing Company.

On the basis of a bilateral agreement we will receive all CDS catalogues in the future.

At the present we have 197 catalogues on magnetic tapes, but we have detailed descriptions for only 139.

After getting new magnetic tapes we make two copies. In order to avoid damage because of the aging of the tapes they are copied every two years.

To provide scientists with better and more specific information, we registered on cards the catalogues available at the moment. This registry is continually extended as new magnetic tapes arrive from the CDS. Every card contains a brief description, that is, not only the CDS-number, but the remarks, contents, number of records, etc... We send this registry to more than 80 different observatories and astronomical institutions in the GDR and in other socialist countries. So far, we have dispatched 139 cards to each address. They also received information about the kind of work which our Data Centre can carry out with the catalogues.

The working group created a system of programming support, that makes it possible to obtain the following information from the CDS catalogues :

1. A listing of a catalogue (only for small catalogues) and a complete copy on magnetic tape

2. A selection of all objects from a catalogue with definite parameters (for example, coordinates, spectral type, colour, radial velocity, etc...)

3. Statistical investigations (for example, frequency of appearance of objects with definite parameter)

4. The calculation of coordinates of celestial bodies for observation epoch with data from a definite catalogue, and so on.

Fortran was used to set up these programs.

During the three years that have passed since the first information on catalogues was dispatched, more then fifthy orders have been received and answered.

Of these, twenty eight orders were for complete copies on magnetic tapes.

In the future, we intend to extend our service for customers, in particular by providing plots.

The work of data collection for the "Bibliographical catalogue of Variable Stars" will continue, and no considerable changes in the work of the group are planned. We hope that our cooperation with the CDS will be in the future as successful as it has been in the past.

IUE DATA DISTRIBUTION

André Heck
Astronomy Division, ESTEC, European Space Agency
ESA Satellite Tracking Station, Apartado 54065, Madrid, Spain

ABSTRACT

The IUE project is described, with special emphasis on the IUE data management and archive retrieval.

This paper briefly describes a more specific data center providing also a service to the astronomical community, or more precisely a group of data centers dedicated to a given instrument.

This instrument is the IUE (International Ultraviolet Explorer) launched on Jan. 26, 1978 and collecting since then spectral images of all kinds of celestial objects in the ultraviolet wavelength range.

The IUE characteristics have been detailed elsewhere (Boggess et al., 1978). We will just remind here that combining the two on-board spectrographs (working respectively in the ∿1150-2000 Å and ∿1800-3000 Å ranges) and the two dispersion modes (low: 6 Å resolution - high: 0.1-0.2 Å resolution), essentially four different types of spectra can be obtained.

IUE is the first completely active "space telescope" with a 24h/day operation much similar to that of ground-based telescopes, with visitors actually coming to the ground stations and participating in the planning and the observational loop - a major difference being that the telescope is not a few meters away in an adjacent dome, but geosynchronously orbiting in space at an average distance of 40,000 km.

Practically all types of objects have been successfully observed with IUE. Already at the XVII IAU General Assembly in Montréal, Meteor, the Assembly Journal, commented that "[IUE was] the first 18-inch telescope in history to provide material for an entire Joint Discussion, and all within 18 months of its first flight".

C. Jaschek and W. Heintz (eds.), Automated Data Retrieval in Astronomy, 29–31.
Copyright © 1982 by D. Reidel Publishing Company.

After this general advertisement for the project, let us see now more concretely what is the IUE output. This is basically a two-dimensional image of 768 pixels over 768 pixels, each pixel with 256 possible intensity levels (8-bit representation). After observing with IUE, the users go home with:

- magnetic tape(s) containing the raw image(s) and various steps of a standard reduction;
- photowrites, two-dimensional photographic representations of the raw image(s) and various steps of the reduction;
- calcomp plots, one-dimensional graphic representations of the gross and net spectra, as well as of the background.

IUE is a collaboration between three agencies which are presently sharing the observing time according to their initial investment in the project: NASA (2/3), SERC (1/6) and ESA (1/6). However the satellite is operated from two stations only: GSFC, run by NASA for NASA observers, and VILSPA, run by ESA for SERC and ESA observers.

Here is a short statistics of the images collected during the three first years of IUE operations (take into account two first months of commissioning and the fact that operations have been gradually speeded up):

Long wavelength spectra (LWR camera)	low dispersion	6643
	high -	3612
Short wavelength spectra (SWP camera)	low dispersion	7397
	high -	3391

Totals:	low dispersion	14040	long wavelength	10255
	high dispersion	7603	short wavelength	11388
		21643		21643

The IUE data management is regulated as follows by the Memoranda of Understanding between the agencies:

- the observers have a reserved use of their data for a period of six months;
- the data are deposited at the data centers (one for each agency: NSSDC for NASA, WDC-C for SERC, and VILSPA for ESA;
- the data are available to the international scientific community (after the reserved period);
- lists of all observations conducted are published (refer to NASA & ESA IUE Newsletters).

An astronomer interested in getting some IUE images (only magnetic tapes are provided in this service) should contact the closest data center and refer to the images by image number. We are now considering to access our archive in the future also by other keys, as object identifiers, object types, coordinates,...

Concludingly, let me just stress that IUE is a foretaste of what will be the Space Telescope to be launched by the Shuttle in 1984-85 and which will carry various auxiliary instruments of different types. I have here a provisional figure of 3 10^9 bits transmitted downwards every day. It is obsviously time to start thinking on the problem of managing, archiving and later distributing these data.

REFERENCE

Boggess et al. 1978, Nature 275, 377.

INFRARED ASTRONOMICAL DATA BASE AND CATALOG OF INFRARED OBSERVATIONS

Marion Schmitz*, Daniel Y. Gezari, Jaylee M. Mead
NASA/Goddard Space Flight Center, Greenbelt, Maryland 20771

Abstract

A computer data base of infrared astronomical observations has been established at NASA/Goddard Space Flight Center. It contains a summary of all infrared (1μm-1000μm) observations of celestial sources outside the solar system, published in the major scientific journals since 1960, as well as the contents of infrared surveys and catalogs. A Catalog of Infrared Observations (CIO) has been developed from the data base in printed and magnetic tape versions. A bibliographic Guide to the Infrared Astronomical Literature, and an Atlas of Infrared Source Names and Positions will be published in conjunction with the catalog. Future plans include development of an interactive data system at Goddard which will give a user direct access to the computerized data.

Infrared Data Base and Literature Search

The Infrared Astronomical Data Base is located at NASA/Goddard Space Flight Center (GSFC). It represents a machine-readable library of infrared (1μm to 1000μm) observational data published in the relevant scientific literature for celestial sources outside the solar system. Catalogs derived from the data base will be published periodically for distribution to interested researchers. To date, over 1100 journal articles and 10 major survey catalogs have been included in this data set, which contains about 70,000 individual observations of 10,000 different infrared sources. Of these, about 8,000 sources have been identified with visible objects and 2,000 do not have known visual counterparts.

The Infrared Astronomical Data Base has been constructed through an extensive search of the appropriate scientific journals and published infrared survey catalogs. This material has been cross-checked with the NASA/GSFC library RECON computer system and the Astronomy and Astrophysics Abstracts under applicable keywords. The infrared literature search presently covers 20 scientific journals for the years 1976-1979

C. Jaschek and W. Heintz (eds.), Automated Data Retrieval in Astronomy, 33–35.
Copyright © 1982 by D. Reidel Publishing Company.

and for several of the most popular journals from 1979-1966. It is expected that completion of the literature search for the present journals as far back as 1960 will increase the number of observations in the data base by about 15%. The data base uses the NASA/GSFC IBM S-360/75 and S-360/91 computers. The format of the current data base is designed to facilitate the data acquisition and verification process, and provides a vehicle for development of future improvements.

Catalog of Infrared Observations

The first product of the infrared data base was the Merged Infrared Catalogue (MIRC) (Schmitz et al. 1978), a compilation of five other infrared catalogs. The 13,588 observations in the MIRC from 2.2μm to 27.4μm are listed in both magnitude and flux units. The next catalog produced will be the Catalog of Infrared Observations (CIO), which contains observational infrared data published in the astronomical literature, in addition to the contents of the MIRC. The CIO contains the following information, as given in the original published reference:

SOURCE NAME
SOURCE POSITION (right ascension and declination at epoch 1950.0)
BEAM SIZE (or aperture)
WAVELENGTH OF OBSERVATION (microns)
INFRARED FLUX (units given in the original reference)
BIBLIOGRAPHIC REFERENCE NUMBER (of the original reference)
POSITION REFERENCE NUMBER (if position not in original reference)
COMMENTS (observational details and object identifications)

If the position of the infrared source is not explicitly given in the original reference, the position was obtained by the editors from other references. The beam size (or aperture) is presented as given in the original reference. The beam size column of the CIO contains a dash if no aperture information is given in the original reference. The beam size is useful in conjunction with observations listed, as an indication of positional accuracy, and as an aid in determining positional coincidences with other sources. The CIO lists the wavelength of each observation in microns, followed by the observed flux at that wavelength, in units as given in the original reference. To preserve the integrity of the data, no attempt has been made to convert the different systems into a standard set of intensity units. The CIO contains additional comments on observational data including spectroscopic information (indicated by the starting and ending wavelengths of the spectrum), source diameter measurements and polarimetry. Upper limit measurements are indicated.

Bibliographic Guide and Atlas of Names and Positions

The Guide to the Infrared Astronomical Literature, included as an appendix to the CIO, indexes the catalog entries to the original references and provides a complete bibliography of the published infrared astronomical literature. The 1,100 infrared journal articles and catalogs are listed in this appendix, with 49 additional references used for position determinations or other data. The bibliographic Guide is arranged chronologically by reference number (which contains the year and month of publication) followed by the authors' last names and initials, journal name or document number, volume, page number, and full title of the reference. The bibliography is also listed alphabetically by author.

The Atlas of Infrared Source Names and Positions is a cross-index of object names, positions, name aliases, and identifications made in the literature or other catalogs. The Atlas is used for locating a source in the CIO by name rather than position, and for identifying observations of the same source which are listed in the catalog under different names. The Atlas is sorted alphabetically by all of the infrared source names in the CIO. Entries from over 100 catalogs are contained in the Atlas of Infrared Source Names and Positions.

Development of the Data Base and Catalogs

Concurrent with the completion of the literature search as far back as 1960 and the expansion of the data base, several improvements in the data base development program are planned. A positional coincidence analysis will be performed to determine which of the observations overlap in the sky. Two-dimensional graphical displays of the infrared data will be generated as overlays for photographic sky surveys. An attempt may be made to reduce the infrared fluxes to a standard system of units. A user-interactive data terminal system will be implemented at Goddard Space Flight Center to give individuals direct access to the computer data base. Results of these efforts may be combined to produce a homogeneous infrared sky catalog of standard source names, positions, and intensities.

Acknowledgments

This work developed out of an infrared catalog program started by Larry Brown in the Laboratory for Extraterrestrial Physics at Goddard, which produced the Merged Infrared Catalog (MIRC). We are grateful to Dr. M. Mumma and Dr. M. Hauser for their continuing support of our infrared catalog and data base research. We thank Dr. T. Nagy and Dr. W. Warren for their helpful contributions. This work is supported by NASA/Goddard Space Flight Center and by NASA contract NAS 5-24360.

References

Schmitz, M., Brown, L. W., Mead, J. M., Nagy, T. A.: 1978, Merged Infrared Catalogue, NASA TM-79683.

*Computer Sciences Corporation, Silver Spring, Maryland 20910

DATA ON TIME AND POLAR MOTION: IMMEDIATE ACCESSIBILITY

Dennis D. McCarthy and Gart Westerhout
U. S. Naval Observatory
Washington, D. C. 20390 U.S.A.

ABSTRACT

Users of time and polar motion information have requirements for data which are somewhat different from those usually considered for astronomical data files. Because of these requirements, it is necessary to provide immediate access to the latest information on Earth orientation and even predictions of future orientation. For other applications where data of the highest accuracy is required with little concern for the delay between observations and the determination of the required information, a more traditional approach may be used. To meet all of these needs a versatile automated data retrieval system is demanded. The U. S. Naval Observatory is now implementing such a system in the area of Earth orientation information and for precise time and time interval applications. Data are available through a series of weekly, monthly, and annual reports, as well as through a digital communications link to Time Service machine-readable files.

INTRODUCTION

The U. S. Naval Observatory provides, as part of its mission, data necessary for accurate navigation on Earth and in space. In this context, time, both atomic and astronomical, and polar motion information are essential. Astronomical time (UT1) and polar motion may be considered as angles which are required to specify completely the orientation of the Earth in a quasi-inertial reference system (Kovalevsky and Mueller, 1981). Astronomical time is a measure of the rotation angle of the Earth while polar motion refers to the changing position of the rotational pole with respect to the Conventional International Origin (CIO) as defined by adopted station coordinates and models. Atomic time defined by the frequency of an energy level transition in the cesium atom serves as a basis for civil time and provides a uniform time scale for practical applications.

C. Jaschek and W. Heintz (eds.), Automated Data Retrieval in Astronomy, 37–44.
Copyright © 1982 by D. Reidel Publishing Company.

Earth orientation information and intercomparison of time scales may be obtained only through observations. In the case of Earth orientation, observations are made to derive the necessary angles. Time and frequency measurements are made to determine accurate time scale data. The information derived from these observations and measurements must be provided to users in a manner suitable to meet their needs.

DATA CONSIDERATIONS

The usefulness of data available to the user may be characterized by a number of parameters. Among these are precision, accuracy, consistency, resolution, and availability. By precision we refer to the formal mean error of the determination as derived from the observations made during one observing session. External precision measures the variations from session to session, and is numerically estimated using the residuals from a smooth curve or the Allan Variance. Precision is measured with respect to an internal reference system. Accuracy, however, incorporates the concept of a "standard" reference system, and is used to characterize the agreement of the derived data with a standard data set. Numerically this may be estimated using the rms of the residuals with respect to a standard data set.

In general, systematic errors may exist in the derived data, and if these can be accurately estimated it is possible to refer the data to the standard reference system. Consistency refers to the stability of the systematic errors in time. A highly consistent system is one in which the systematic errors do not change over extended periods of time. Resolution in this context refers to how often independent estimates of the derived data are available. High resolution implies a rather short observing session in order to obtain useful data. Availability refers to the time interval between the end of the observing session and the time when the derived data are available. Those making use of the derived data must evaluate their needs in terms of these quantities to determine the usefulness of various possible sets for their particular applications. For example, high accuracy may have to be compromised in favor of high resolution or immediate availability.

USER REQUIREMENTS

Ideally, it would be desirable to have high accuracy data with the highest resolution available almost immediately after observing. Realistically, compromises must be made. Thus it seems appropriate that requirements for time and polar motion information may be classified into three categories.

"Historical" data of the highest possible accuracy and consistency are needed by those engaged in scientific research and for geodetic purposes. While it may be desirable to have the data as soon as possible, users of historical data may not be concerned with immediate availability if it is only possible at the expense of reduced accuracy or consistency.

"Current" data are required by users needing high precision and almost immediate availability. They are willing to sacrifice the consistency and accuracy of the data to some extent in order to meet their need for the most current information available. Those involved with navigation frequently fit into this area.

"Future" data are required for those applications where predictions are required. For example, if predictions of some phenomenon are demanded in which time and polar motion play a part, these users must have projected values with the highest possible precision and consistency.

These three types of user requirements, in turn, demand a versatile system for data retrieval. For the user of historical data, reports issued annually may be sufficient. Clearly this would not be useful to those who need predictions of time and polar motion updated on an immediate basis.

AVAILABLE DATA

Time and polar motion information are currently available from a number of sources including international services. The International Latitude Service (ILS) organized in 1899 has provided polar coordinates since that time. These have been obtained using the observations of five visual zenith telescopes located on the parallel of 39° N. While these locations have served as the definition of the CIO, the accuracy of the data derived from these observations has been superseded by the use of more modern techniques and methods. ILS data are available generally after a processing delay of a few months.

Since 1962, the International Polar Motion Service (IPMS) has provided polar coordinates derived from observations made with approximately sixty classical astronomical instruments located around the world. These data are available some months after the actual observations.

The Bureau International de l'Heure (BIH) provides time information as well as polar coordinates. Both astronomical and atomic time are coordinated through this international service. The polar coordinates are determined making use of information from the Doppler tracking of artificial satellites as well as the results from a large number of classical astronomical instruments. The data are generally available one to two months after the observations.

The BIH also maintains a Rapid Service which provides Earth orientation information based on the observations of a subset of the previously mentioned classical instruments. The data are available from one to ten days following observations at an accuracy somewhat lower than that associated with the regular BIH data.

In addition to these data sets, a number of independent data sets are available containing data from various observatories. For example, the U. S. Naval Observatory has a continuous record of the of variation of astronomical latitude at Washington since 1915 and UT1-UTC since 1955. These data are based on observations made with a photographic zenith tube. Similar information is available from other observatories, including the Royal Greenwich Observatory, Paris, Hamburg, Ottawa, and Tokyo. Polar coordinates derived from the tracking of Doppler navigational satellites are also available from the U. S. Defense Mapping Agency starting in the late 1960's. Recently, modern techniques such as laser ranging to artificial Earth satellites and to the Moon, and radio interferometry, have provided additional sets based on completely different observational methods. Possible systematic errors involved in each of these methods remain to be investigated, but these observations show a new promise for increased accuracy and consistency.

For predictions of future Earth orientation parameters, the U. S. Naval Observatory also provides routine weekly updated predictions of astronomical data. They are available through weekly publications and by means of a digital communications link. This latter service (see Time Service Announcement Series 14, No. 27. 1980) provides the latest information available in the area of precise time and time interval as well as the predictions of Earth orientation.

SUMMARY OF U. S. NAVAL OBSERVATORY TIME SERVICE DATA

Appendix A lists the Time Service Publications currently available on a weekly, monthly or annual basis. As mentioned above, a digital communications link is also available for immediate automatic retrieval of data. Among some of the data files available through this service are the notes of the most recent Series 4 publication (see Appendix A), up to the last ten Series 5 messages, predictions of polar coordinates and UT1, and preliminary values of the difference between the Master Clock and UTC(USNO). Also included in the approximately forty files available are general messages of interest to the user as well as information on the status of OMEGA, Loran-C, and GPS. The most recent monitoring information from GPS, OMEGA, Loran-C, TRANSIT, and television timing is also available. A time signal is provided which can be used to set a computer on time if the communications link is called by the user's computer and an appropriate program to set the system clock is available. The precision is currently limited to ± 10 milli—seconds. The data of all these files and time signal are available immediately to the user with a suitable computer terminal. Thus the digital communications link is designed to meet the needs of those users for immediate information. Appendix B contains a partial explanation of the use of the communications link. Further information can be obtained by writing to: Director, Time Service Division. U. S. Naval Observatory. Washington, D. C. 20390 U.S.A.

FUTURE PLANS OF THE U. S. NAVAL OBSERVATORY

To meet user requirements the U. S. Naval Observatory is currently implementing a system of astronomical time and polar motion data based on the use of classical observations as well as the results available from the modern techniques (McCarthy. 1981). The latter methods include laser ranging to the Moon and to artificial satellites and the connected element interferometer at Green Bank. Predictions of future Earth rotation parameters will be based on this set of data. It is expected that all data types (historical, current, and future) will be available to the user immediately in machine-readable form through the digital communications link. The user will need to specify only the date(s) for which the information is required and the data will be retrieved along with the corresponding estimates of precision.

Already existing files dealing with precise time and time interval information will be expanded as needed to suit user requirements. For example, it is expected that the current limitation in precision of the time signal provided by the digital communications link will soon be improved to meet user needs.

CONCLUSION

As future requirements for time and polar motion information become more stringent both in terms of accuracy and availability it will become more important to assure that data files can be retrieved immediately by users. The U. S. Naval Observatory is already providing such a system for users of precise time and time interval information as well as those requiring predictions of Earth orientation. It is planned that this system will be expanded to meet already existing demands as well as future requirements.

APPENDIX A. LIST OF U. S. NAVAL OBSERVATORY TIME SERVICE PUBLICATIONS

Series 1 WORLDWIDE PRIMARY TIME and FREQUENCY VLF and HF TRANSMISSIONS. Includes call sign, geographic location, frequencies, radiated power, times of broadcast, etc., of radio transmissions suitable for precise time measurement. Contains sections pertaining to U. S. Navy time and frequency transmissions, Loran-C, OMEGA, National Bureau of Standards (NBS) and other time signals. (Issued as necessary.)

Series 4 DAILY PHASE VALUES and TIME DIFFERENCES. Lists observed phase and/or time difference between VLF, LF, OMEGA, television, portable clock measurements, Loran-C stations and the U. S. Naval Observatory (USNO) master clock, UTC(USNO.MC). Propagation disturbances and notices of interest for precision timekeeping are also given. (Issued weekly.)

Series 5 USNO PHASE VALUES/TELETYPE MESSAGE. Lists information described in Series 4 as it becomes available. TWX message for U. S. Government addresses only after submission of written justification to the Superintendent.

Series 6 A.1 - UT1 DATA. Lists daily values of polar coordinates. correction for seasonal variation. and A.1-UT1 as observed at USNO and Naval Observatory Time Service Substation (NOTSS), Richmond, Florida. The astronomical latitude as observed at each station is also given. (Issued monthly.)

Series 7 PRELIMINARY TIMES and COORDINATES of the POLE. Lists general time scale information, values of UT1-UTC predicted two weeks in advance, the Bureau International de l'Heure (BIH) values of UT1-UTC and polar coordinates. Doppler Polar Motion Service, IPMS and ILS polar coordinates are also listed as they become available. (Issued weekly.)

Series 8 TIMES of COINCIDENCE (NULL) EPHEMERIS TABLES for TELEVISION. At present these tables are applicable only for WTTG Washington, D. C. and KTTV Los Angeles, CA. They may be of interest in countries operating on the NTSC system. (Issued annually).

Series 9 GENERAL INFORMATION for LORAN-C. Individual tables are issued for the master station of each Loran-C chain. (Issued annually.)

Series 10 ASTRONOMICAL PROGRAMS. Includes information pertaining to results, catalogs, papers, etc., concerning the Photographic Zenith Tube (PZT), Danjon Astrolabe, and Dual-Rate Moon Position Camera. (Issued as available.)

Frequently this information will be released as a Time Service Announcement Series 14.

Series 11 TIME SERVICE REPORT. List general timing information and time differences between coordinated stations and the UTC time system; adopted differences UT1-UTC and A.1-UT1; UTC(USNO MEAN)-UTC(USNO MC); UTC(USNO MC)-UTC(BIH); astronomical latitude and UT1 as observed at USNO and NOTSS; polar coordinates and corrections for seasonal and polar (longitude) variations. (Issued annually.)

Series 13 PRECISE TIME and TIME INTERVAL PLANNING MEETING. Includes announcement of the meeting held each December in Washington, D. C., the call for papers, and the preliminary program.

Series 14 TIME SERVICE ANNOUNCEMENTS. Includes general information pertaining to time determination, measurement, and dissemination. (Issued as required.)

Series 15 BUREAU INTERNATIONAL de l'HEURE CIRCULAR D. Lists Universal Time and coordinates of the pole; emission time of time signals; Universal Time (Coordinated) from Loran-C and television pulse receptions and independent local atomic time scales (AT_i). This publication is distributed to U. S. addresses only. (Issued monthly.)

Series 16 PRECISE TIME TRANSFER REPORT. Lists the time difference UTC(USNO MC)-UTC (reference clock), adjustments to reference clocks and portable clock measurements. The time difference is obtained via communication satellite time transfer, television and/or Loran-C receptions. (Issued each 20 days.)

Series 17 TRANSIT SATELLITE REPORT. Lists the difference UTC (satellite clock) - UTC (USNO MC) and the frequency offset for each of the operational satellites. The information published is received from Naval Astronautics Group, Pt. Mugu, CA. (Issued weekly.)

Series 18 WORLD DATA CENTER-A, ROTATION of the EARTH, INFORMATION BULLETIN. Lists proposed times for experiments utilizing modern techniques, such as radio interferometry and laser ranging. When available, preliminary values for polar motion and Universal time obtained from these techniques are given. (Issued as available.)

NOTES:

(1) Past issues of Series 4 are available on microfiche.
(2) Series 6 is available in machine-readable form at the usual exchange ratio of three-to-one.

APPENDIX B. EXPLANATIONS FOR USE OF THE USNO REMOTE DATA ACCESS SYSTEM

You have available a number of info services which can be called by sending a key sequence as given in Table of Codes, terminated by CR. Example: To get Table of Codes type in @TCO (and carriage return).

Use full duplex, 300 or 1200 BAUD, even parity, ASCII seven level, 10 bits per character with start, parity and stop bit. Please note that no spaces are allowed within the Command Code and that only upper case characters will be accepted for that purpose. Your messages, however, may contain any ASCII character except, of course, Control D (EOT), line feed and carriage return which are reserved for their respective purposes.

You may leave a message for NAVOBSY whenever you get the USNO info prompt *. Terminate your call with CNTL D (EOT). This is important to make the program immediately available again. For interrupting a long file being read to you, send "BREAK" (CNTRL 2 on some TI Terminals). The requested identification can be of any length since we have a logfile on which everything incoming is kept.

You will obtain best results if you do not call during peak traffic, i.e., not between 9-12 a.m. and 1:30-5 p.m. EST, particularly on AUTOVON. If you should get poor results ("garbage") try to terminate and call again. Line quality can change greatly from one call to the next and tends to deteriorate during peak traffic time. Also make sure that your handset is seated firmly in the terminal rubber receptacles and that your microphone is working properly. Difficulties in access may be cured with slight transmit level adjustment on your terminal. Some of our callers have also noticed that trouble may exist on a particular extension and in this case it is your local line which has to be checked. The Phone Company also offers a "WEAK VOICE" amplifier for less than one dollar per month which will boost your modem level to compensate for a lossy line to your central office. Lines with too many extensions are more likely to be troublesome. Simply try a different telephone.

Since 300BD is different in respect to marginal conditions compared to 1200BD one should try both speeds if difficulties persist. If you use a computer to talk to us then please remember that our identification will come on automatically. However, there is a greater delay on 1200BD before we start transmitting compared to 300BD. It is therefore not necessary to send interrupts. Unless sent as breaks for interrupting reading of a long file, these interrupts will not speed up your access since the interface is insensitive to interrupts after your call has been answered.

Our normal operation uses ECHO. i.e., your terminal in full duplex writes what it gets back from us. For certain computer traffic, however, it may be advisable to turn off the ECHO. While we have three dummy (NULL) characters at 300BD to allow slow printers to follow, we do not have them at 1200BD. However, this can be changed by operator command. This command changes the operation into its opposite (but not back).

REFERENCES

Kovalevsky, J., Muller, I. I., 1981. "Comments on Conventional Terrestrial and Quasi-inertial Reference Systems" in Reference Coordinate Systems for Earth Dynamics, E. M. Gaposchkin and B. Kolaczek (eds.), D. Reidel, Dordrecht, pp. 375-384.

McCarthy, D. D., 1981, "On the Adoption of a Terrestrial Reference Frame" in Reference Coordinate Systems for Earth Dynamics, E. M. Gaposchkin and B. Kolaczek (eds), D. Reidel, Dordrecht, pp. 145-154.

Time Service Announcement Series 14, No. 27, 1980, available from U.S. Naval Observatory Time Service Division, Washington. D.C. 20390 U.S.A.

THE NATIONAL SPACE SCIENCE DATA CENTER (NSSDC) AND THE WORLD DATA CENTER A FOR ROCKETS AND SATELLITES (WDC-A-R&S): THEIR ROLE IN X-RAY ASTRONOMY DATA

M. C. Locke
National Space Science Data Center
NASA/Goddard Space Flight Center
Greenbelt, Maryland USA

ABSTRACT

The combined role of NSSDC and WDC-A-R&S in collecting, preserving and disseminating data from 10 years of X-ray observations is presented along with a brief history of the organizations.

NSSDC AND WDC-A-R&S

The National Space Science Data Center was officially established in November 1965, as part of Goddard Space Flight Center, Greenbelt, Maryland, USA, to provide the means for further analysis and dissemination of space science data beyond that provided by the analyses, presentations and publications of the principal investigators and their co-workers. Data deposited in NSSDC are made available to scientists outside the USA through the co-located World Data Center A for Rockets and Satellites.

In order to fulfill its mission, the NSSDC acts as a central repository for the data from space science flight experiments in at least two general forms. Reduced data records, which are usually prepared by a compaction, editing, and merging operation performed by the principal investigator, form the primary data base for the NSSDC. Data in this form contain all of the the basic usable information obtained from the experiment and include the instrument responses measured as functions of time along with the appropriate ephemeris and spacecraft information necessary to analyze the data in an independent fashion. Hopefully, in this way there will be a minimum of information removed from the data by analysis techniques which the principal investigator favors for his own interpretation of the results. A second form of the data which must be collected is the final analyzed data which the principal investigator designates as the best to display the scientific results of the experiment. This form may include charts, graphs, and photographs, and tables which are the

C. Jaschek and W. Heintz (eds.), Automated Data Retrieval in Astronomy, 45–47.
Copyright © 1982 by D. Reidel Publishing Company.

results of data processing techniques employed by the principal investigator but are too numerous to be contained in the published works in their entirety.

Primarily the NSSDC has used manual or semi-automated techniques for storing, retrieving, and reproducing the data. These techniques include microfilming or microfiching data to reduce storage space, using microform and photographic copy devices, and regenerating magnetic tape through standard computer copying routines. In addition an attempt is made to assure later investigators independent use of the data by providing additional information and services such as descriptions of the instrument, calibration history, references to published scientific results and all other pertinent bibliographic materials.

However, the various measurements of space astronomy parameters are made in numerous ways and can cover widely different energy-time regions, often making it difficult for an individual investigator to utilize the data from different experiments. Yet in order to fully characterize the nature of a particular astronomical phenomena it may be necessary to compare the results from many satellites. This procedure can involve a tremendous amount of data, the processing of which may not be feasible for individuals or groups. Data compilations are one solution and are consistent with the purposes of further use and dissemination of space science data. Toward this goal, NSSDC is currently involved in the acquisition of data from past X-ray astronomy satellite experiments.

X-RAY ASTRONOMY DATA

Although extrasolar X-ray sources were first discovered in 1962, the full scope of X-ray astronomy began with extended observations by the satellite Uhuru (SAS-A) which was launched in December 1970. During the past decade since Uhuru's launch more than 11 major satellite missions have been flown with instruments dedicated in whole or in part to exploring the X-ray sky. The discoveries made by these missions have clearly established that X-ray observations are an essential tool in the study of the dynamics and evolution of many cosmological objects. New missions and advances in technology will not make these data obsolete but to the contrary, expanding knowledge gives new insight and often new meaning to early observational phenomena. Not every mission could cover all energy ranges nor take all the different types of measurements; however, the combined data available from these early experiments spans not only a full decade in time (1970-1980) but over five decades of the electromagnetic spectrum (10^2-10^7eV). The overlap of these missions in both time and energy presents an impressive possibility of compiling timelines that for many sources would reflect 10 years of activity in several energy ranges (Table 1 indicates missions from which data are currently becoming available). There are many unanswered questions in astronomy that require no new technology or sophisticated instruments, but are best answered by having observations with continuity over long time spans.

Data compilations are precisely the type of intermediary function that can be performed by a central data repository such as NSSDC. The reduced data records from many experiments can be sorted and compiled by parameters such as position, time, energy or any other area of interest and then made available with appropriate documentation to requesters for further analysis. This eliminates the need to copy, ship, and process large volumes of tapes or microform in order to sort out specific information. It is also a way to provide non-X-ray astronomers with the available data in the form most complementary to their own measurements.

Services can also be rendered for the previously analyzed data that would further its use. For example, the many published source catalogs and positions can be compiled and made available in machine readable form; bibliographies of published results can be assembled; and cross references can be generated for data available in assorted forms such as plots, charts, tables, or images.

By using the facilities of NSSDC and WDC-A-R&S in this manner, the full potential of the information gathered from these and other space science investigations can be realized and made available to not just a select few, but to the entire world scientific community.

TABLE 1. X-RAY ASTRONOMY MISSIONS 1970-1980

SPACECRAFT	LIFETIME	ENERGY RANGE
UHURU	DEC. 1970 - JAN 1975	$1x10^3$ - $2x10^4$ eV
OSO 7	SEPT. 1971 - JULY 1974	$1x10^3$ - $6x10^5$ eV
OAO 3	AUG. 1972 - DEC. 1980	$1x10^2$ - $8x10^3$ eV
ANS	AUG. 1974 - JULY 1976	$2x10^2$ - $4x10^4$ eV
UK 5	OCT. 1974 - MAR. 1980	$3x10^2$ - $3x10^4$ eV
SAS-C	MAY 1975 - APR. 1979	$2x10^2$ - $6x10^4$ eV
OSO 8	JUNE 1975 - OCT. 1978	$2x10^3$ - $1x10^6$ eV
HEAO 1	AUG. 1977 - JAN. 1979	$1x10^2$ - $1x10^7$ eV
HEAO 2	NOV. 1978 - APR. 1981	$1x10^2$ - $2x10^4$ eV
UK 6	JUNE 1979 - PRESENT	$1x10^2$ - $3x10^4$ eV
HEAO 3	SEPT. 1979 - MAY 1981	$6x10^4$ - $1x10^7$ eV

THE STELLAR DATA FILE AT DEARBORN OBSERVATORY

William Buscombe
Northwestern University

The majority of entries for <u>MK Spectral Classifications</u> in the Kennedy & Buscombe (1974) catalogue referred to stars brighter than 9th magnitude with H. D. numbers. Accordingly, the file was arranged in sequence corresponding to right ascension for 1900, although the co-ordinates were actually listed for the epoch 2000, in accordance with an earlier precept established in the IAU.

When similar data were being assembled for the <u>Third General Catalogue</u> (Buscombe, 1977), many fainter objects were included, such as members of the Magellanic Clouds, certain types of variable stars, white dwarfs. More than half the entries have no H.D. number, and unfortunately some errors in sequencing (and even precession!) occurred. For this file, a particular effort was made to include classified stars with their UBV photometry, which have been confirmed as members of star clusters and physical binary systems.

The worldwide acceptance of these paper-bound books has led us to offer machine-readable versions on magnetic tape, transcribed to order for several institutions. Continual updating of the card files to include additional data and identifications is performed with the assistance of one part-time employee, currently Mark Shickel, an undergraduate student, who has taken increasing responsibility in handling the computer facilities. A merged file of the 1974 and 1977 material, with many amended entries, is available. Indications of peculiarities in chemical abundances are included where appropriate.

To facilitate future filing, the cards of the <u>Fourth General Catalogue</u> (Buscombe 1980) are arranged in order of R. A. (2000), so that the order of H.D. numbers is not maintained. Many identifications of stellar x-ray and infrared sources were included. The labour of maintaining files of references has been eliminated, because these are performed extensively at Meudon, France, and the bibliography is accessible in the semiannual <u>Astronomy and Astrophysics Abstracts</u>. Similarly, the lists of abbreviations for key initials of lists of faint stars are being organized at Meudon and Strasbourg, but the data

C. Jaschek and W. Heintz (eds.), Automated Data Retrieval in Astronomy, 49–50.
Copyright © 1982 by D. Reidel Publishing Company.

compilations which have been distributed on microfiche do not include as many faint objects as our catalogues. Some data files are more comprehensive, but we have limited each entry to an 80-column card. We try to maintain an accuracy within 0.3 min. of R.A. and 3' of Dec.

As the classification system has been adapted for use in many observatories with quite diverse equipment, secondary standards have been proposed to supplement the objects which originally defined the MK system. About 600 of these, selected from the recent atlases by Morgan, Keenan, Jaschek, and Yamashita, are listed on the opening pages of our forthcoming Fifth General Catalogue. The next section of about 60 pages assembles all available data for classified member stars of the Small and Large Magellanic Clouds. Earlier information from Cannon, Henize, Feast, Thackeray, Wesselink, Westerlund, Stock, Sanduleak, Humphreys, Osmer, Fehrenbach, Duflot, Martin, Brunet, Azzopardi, Breysacher, Ardeberg, Isserstedt, Bigay, Florsch, Gascoigne, Lloyd Evans, Hodge, Essser, and others, is consolidated (with slight corrections to positions, where necessary) in single sequence. The final group of 15,000 entries continues the compilation, mostly from literature of the last five years, especially based on Bartaya's classifications in the Northern Kapteyn Selected areas, certain stars of early type in the Southern Milky Way, and additional infrared sources. Wherever possible, UBV photometry from the Strasbourg microfiche is included, as well as notes on duplicity. The far left-hand columns indicate how many stellar images contribute to the quoted spectrum, and the right-hand columns, how many are included in the diaphragm for photometry (if known).

B. DATA NETWORKS

EURONET-DIANE, PREMIER RESEAU EUROPEEN DE TRANSFERT DE L'INFORMATION.

Yvonne SALLE
Equipe de lancement Euronet-DIANE, Luxembourg.

Euronet is one of the world's most recent networks created for data transmission. Sponsored by the Commission of the European Communities, Euronet was officially opened for traffic, at the beginning of 1980, in the buildings of the European Parliament in Strasbourg.

Purpose: To promote the transfer of scientific and technical information within Europe, and to develop the European information industry by encouraging the creation and availability of databases and databanks.

Partners: Host services offering on-line access to databases.
P.T.T.'s of the 10 member countries of the EEC, (and more recently Switzerland), providing the packet-switched network.
EEC who sponsored the project and continues to monitor its activity and development.

Techniques: Packet-switching technology for the transmission of data.

Characteristics: - distant independent tariff

Services:
- 300 data bases in all subjects of science, technology, economics, law: around 60 million information , such as geophysics...
- common command language (C.C.L.),
- on-line ordering of documents,
- automatic translation,
- third party traffic,
- compatibility with other networks (e.g. Transpac, Datex-P...)

Les organisateurs de ce colloque m'ont fait l'honneur de me confier la tâche d'introduire la session consacrée aux "réseaux de données en astronomie", et de présenter Euronet. Euronet n'étant pas un réseau spécialement appliqué à l'astronomie, et n'étant pas moi-même spécialiste de l'astronomie, mon intervention vous paraîtra peut être un peu paradoxale.

C. Jaschek and W. Heintz (eds.), Automated Data Retrieval in Astronomy, 53–61.
Copyright © 1982 by D. Reidel Publishing Company.

Je vais donc essayer, en présentant Euronet, de le situer parmi les grands réseaux internationaux afin de mettre en lumière ce que les technologies nouvelles de télécommunication peuvent apporter à la solution des problèmes qui se posent aux spécialistes que vous êtes, en matière d'information. Ceci est d'autant plus important que l'astronomie est, plus que beaucoup d'autres, une science qui ignore les limites géographiques qui divisent le globe et que votre communauté scientifique a, de ce fait, toujours travaillé au-delà des frontières: je pense notamment au rayonnement des recherches des astronomes de l'Antiquité ou, plus près de nous, d'Oulou Beg, de Copernic...alors que les voyages étaient hasardeux sinon impossibles.

L'assemblée réunie aujourd'hui prouve que les problèmes de communication sont largement résolus. Cependant, les dernières années ont vu l'apparition de nouvelles technologies qui ont permis la création de réseaux pour la transmission des informations en général, réseaux qui ont pris le nom de réseaux de transmission de données. Le terme "réseaux de transmission de données" recouvre plusieurs types de réseaux affectés aux transferts d'information. Réseaux privés ou réseaux publics, leur technologie, leur champ géographique sont très variables. Les premiers d'entre eux ont déjà quelque 10 ans d'existence et sont nés aux Etats-Unis. Ils avaient pour objet de relier deux ou plusieurs ordinateurs entre eux: leurs caractéristiques répondaient donc aux cas particuliers représentés par ces ordinateurs.

Les réseaux plus récents se sont fixés pour objectif de permettre l'accès à l'information d'un nombre très important d'utilisateurs. A la notion d'échange d'information entre deux ordinateurs, on a substitué en quelque sorte, celle de répartition d'informations à partir d'un ordinateur appelé serveur à des utilisateurs munis de terminaux. On assiste donc à une hiérarchisation des rôles: des fournisseurs d'information préparent les fichiers - bases et banques de données - dont ils confient la diffusion à des ordinateurs-serveurs qui les mettent à la disposition des utilisateurs. Avec leur terminal, les utilisateurs peuvent acquérir cette information quand ils en ont besoin, en utilisant un certain nombre de codes qui constituent le "langage de commande" propre au serveur. Le dialogue entre utilisateurs et serveur se réduit à l'usage de ces codes, outil de sélection des informations qui, à la fin du dialogue, sont délivrées par l'ordinateur.

La vulgarisation de la diffusion de l'information est devenue possible en raison de:
- la diminution des coûts des terminaux et c'est le cas des terminaux dits compatibles télétype" (TTY);
- la réduction des coûts de télécommunication, c'est-à-dire de transfert de l'information; c'est ce qu'a permis la technique dite de la "commutation de paquets".

Depuis quelques années, tous les réseaux destinés au transfert de l'information vers le public ont adopté la technologie de la "commutation de paquets" qui permet à plusieurs utilisateurs de faire transiter leur message simultanément sur la même ligne téléphonique et, en conséquence, de diminuer les coûts de transfert.

Les réseaux Télénet et Tymnet aux Etats-Unis, Datapac au Canada, Transpac en France, Datex-P en Allemagne...et Euronet ont adopté cette technologie.

I. Le réseau Euronet-DIANE

Euronet est le premier réseau européen de transmission de données.

Il est né d'une volonté politique qui s'est exprimée au sein de la Commission des Communautés Européennes (CCE) et plus particulière- de la Direction Générale XIII et du Comité pour l'Information et la Documentation Scientifique et Technique (CIDST) dans le but de résoudre un certain nombre de problèmes qui se posaient au monde scientifique.

Parmi les nombreux facteurs qui ont concouru à la décision de créer Euronet-DIANE, on peut distinguer deux groupes importants:

1) tout d'abord des facteurs que l'on pourrait qualifier de techniques:
   - le nombre de chercheurs, d'ingénieurs, de spécialistes dans l'Europe du Marché Commun: il est évalué à 12 millions de personnes, qui utilisent et fournissent régulièrement des informations au cours de leur carrière;
   - la masse grandissante de l'information dans ce seul domaine: au niveau mondial, elle est évaluée à 2 millions de contributions originales, chaque année;
   - la maîtrise, depuis plus de dix ans, des techniques de stockage de l'information en particulier avec le stockage en mémoires magnétiques;
   - la maîtrise de la recherche individuelle et immédiate de l'information parmi plusieurs millions de références stockées;
   - la maîtrise des techniques de diffusion et de circulation de l'information stockée en ordinateur: c'est l'accès en conversationnel à l'ordinateur par l'utilisateur lui-même, au moyen d'un terminal peu coûteux relié à un réseau de télécommunication spécialisé.

2) les facteurs que l'on peut qualifier de politiques et d'économiques:
   - les pays de la CEE comprennent environ 256 millions d'habitants et leur économie est largement développée, ce qui constitue, a priori, un marché pour l'information scientifique et technique;
   - ces pays, dans l'ensemble, à court de matières premières, sont confrontés à une conjoncture économique de plus en plus compétitive; la Communauté économique européenne ne peut progresser qu'en développant soigneusement les ressources dont elle dispose, notamment, les ressources intellectuelles;
   - le volume des informations scientifiques et techniques ou autres, actuellement disponibles ou en cours de rassemblement à travers l'Europe constitue une matière première très riche, dotée d'un

immense potentiel d'exploitation. Pour que l'on en tire le meilleur parti, cette information doit être efficacement stockée, gérée et diffusée: les techniques citées plus haut le permettent.

A la fin des années 60, tous les éléments nécessaires à la réussite d'un réseau consacré à la circulation de l'information scientifique et technique étaient dont réunis; le rôle de la CEE a consisté à rassembler les différents groupes ou organismes intéressés à sa réalisation, à coordonner leurs recherches et leurs activités et à fournir une large contribution financière pour la réalisation finale du réseau Euronet. Sans cette initiative de la CCE, on peut présumer que des réseaux nationaux privés et publics se seraient créés progressivement dans les différents pays mais que leur interconnection se serait révélée difficile en raison notamment de l'absence d'un accord général entre pays et de normes internationales préexistantes.

Euronet a été inauguré le 13 février 1980 à Strasbourg. Près de neuf années se sont donc écoulées entre le jour où le Conseil des Ministres de la CEE approuvait une résolution définissant une politique de base pour la coordination des activités dans le domaine de l'information scientifique et technique et celui où le réseau européen entrait en fonctionnement. Ces neuf années ont été nécessaires pour adopter le principe d'un réseau, en définir les normes et le mettre en oeuvre.

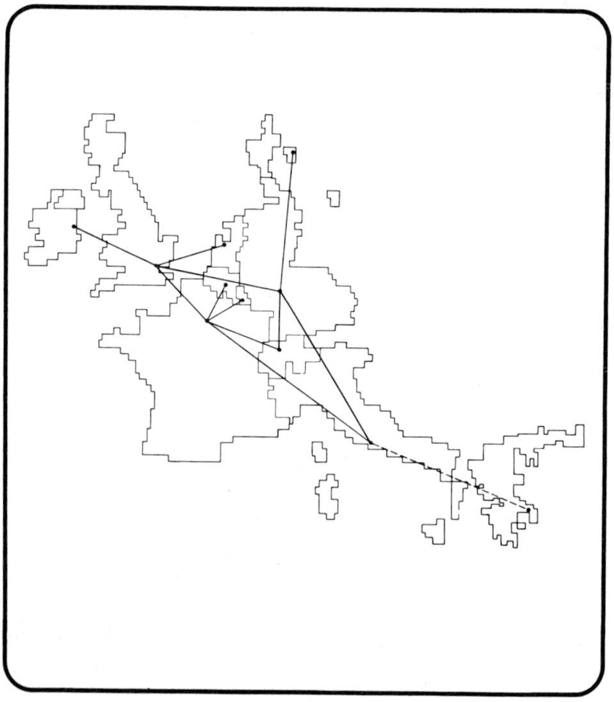

Au plan technique, le réseau EURONET est un réseau de transmission de données par commutation de paquets: il est identique en ce qui concerne la technologie, au réseau français TRANSPAC et il répond aux normes du CCITT (Comité Consultatif International pour le Télégraphe et le Téléphone). Ceci assure sa compatibilité avec les autres réseaux qui de plus en plus adopteront ces normes.

Il comporte:

- cinq noeuds situés respectivement à Francfort, Londres, Paris, Rome et Zurich: la vitesse de circulation de l'information entre les noeuds est de 48.000 bits/sec.;
- dix concentrateurs situés dans les cinq villes précédentes et dans les capitales des autres pays membres soit Amsterdam, Bruxelles, Copenhague, Dublin et Luxembourg.

Chaque pays dispose donc d'un accès sur son propre territoire. L'extension du réseau, à la Suisse, a été réalisée à la suite d'un accord conclu entre ce pays et la CCE. Son extension vers la Grèce, nouveau pays membre de la Communauté, est en préparation.

Les ordinateurs-serveurs qui diffusent l'information sont connectés à EURONET au moyen d'une ligne téléphonique spécialisée qui les raccorde au multiplexeur (PSE)le plus proche. Quant aux utilisateurs qui souhaitent appeler l'un ou l'autre des serveurs, ils connectent d'abord leur terminal au multiplexeur de leur pays au moyen d'un appareil et du réseau téléphoniques classiques.

EURONET a été conçu pour permettre la connexion d'une très large gamme d'ordinateurs; actuellement, les serveurs déjà connectés travaillent notamment avec les ordinateurs suivants: CII, IBM, ICL, ITEL, SIEMENS, UNIVAC...

Il en est de même en ce qui concerne les terminaux: presque tous les terminaux "compatibles télétypes" peuvent être utilisés pour accéder à EURONET par l'intermédiaire du réseau téléphonique commuté et d'un coupleur acoustique, ou d'um modem.

Une autre caractéristique - la plus remarquable au regard des utilisateurs - réside dans la structure des coûts de liaison. Afin de répondre aux voeux du Parlement et du Conseil des Ministres Européens, à savoir, faciliter l'accès des petites et moyennes entreprises à l'information, la tarification mise au point est indépendante de la distance, ce qui garantit l'égalité d'accès à tous, dans l'ensemble des pays de la Communauté. Le prix est donc calculé généralement, en fonction du volume des données transmises et de la durée de la transmission. Ce prix représente une économie de 60 à 80% par rapport aux tarifs internationaux de télécommunications.

A la différence des réseaux principaux existants consacrés aux transferts d'information scientifique et technique qui sont des réseaux privés, les services EURONET-DIANE présentent un caractère coopératif.

C'est, en quelque sorte, une oeuvre collective qui réunit, autour de la Commission des Communautés Européennes:

- les administrations des PTT des pays participants, responsables du réseau de télécommunication lui-même, c'est-à-dire de son fonctionnement, de l'entretien et du développement des installations, de la gestion des abonnés;
- les ordinateurs-serveurs, responsables individuellement de leurs services auprès de leur propre clientèle et qui forment le groupe des serveurs "DIANE". Les serveurs sont indépendants les uns des autres et leurs statuts très différents: l'un d'eux, rattaché à l'Agence Spatiale Européenne, est réellement international; d'autres dépendent de sociétés privées; la plupart,dans une mesure variable, relèvent d'organismes publics ou nationalisés, ou bénéficient de subventions. Ils choisissent également, d'une manière totalement indépendante, leur lieu d'implantation et les bases de données qu'ils vont mettre sur le marché.
- les utilisateurs, réunis en un "forum des utilisateurs" ce qui leur permet de faire connaître leurs souhaits ou leurs desiderata au Groupe des serveurs ou aux PTT.

Le CIDST assure des tâches de coordination et soutient la mise en oeuvre du projet.

II. Les services d'Euronet DIANE

a) Les services sont tout d'abord les bases et banques de données proposées par les serveurs. Leur nombre n'a cessé de croître depuis la création du réseau de sorte que les chiffres mentionnés aujourd'hui deviendront rapidement caducs. Ainsi, le jour de l'inauguration, sept serveurs seulement étaient effectivement raccordés au réseau et diffusaient environ 70 bases ou banques de données; un mois plus tard, le chiffre des serveurs était de douze et le nombre de fichiers disponibles était supérieur à 100. Aujourd'hui, moins d'un an et demi plus tard, les serveurs sont au nombre de 36 et l'on compte près de 300 fichiers, soit plus de 60 millions d'informations. La majorité de ces informations est, encore actuellement constituée de références bibliographiques. La répartition des fichiers par domaine d'activité se présente de la manière suivante:

- scientifique et technique (incluant biologie médecine et brevets)	46 %
- affaires, aspects économiques (y compris les séries statistiques)	23 %
- domaine juridique	16,5 %
- fichiers multidisciplinaires	7,5 %
- sciences humaines et sociales	3 %
- divers	3 %

Il convient de préciser que les informations diffusées au moyen du réseau Euronet, par les serveurs du Groupe DIANE, ne sont pas uniquement d'origine européenne: à côté de fichiers créés dans les pays du Marché Commun on trouve, notamment, les grandes bases constituées aux Etats-Unis à partir d'informations publiées dans le monde entier. Toutefois, la création d'Euronet a constitué un stimulant à la diffusion de nouvelles bases et banques de données européennes dans de nombreux domaines et, particulièrement, dans le domaine de l'économie et des affaires: en cela, le souhait initial des promoteurs du réseau de voir se créer une industrie européenne de l'information est en passe de se réaliser.

b) Pour compléter les efforts faits en vue de développer l'accès à l'information, outre l'accroissement de l'offre, un certain nombre de services annexes ont été créés sous l'égide de la CCE. L'un d'eux est constitué par le "langage commun de commande". Ainsi qu'il a été indiqué plus haut, les utilisateurs emploient une série de codes pour sélectionner l'information dont ils ont besoin; or ces codes, appelés généralement "langage de commande" varient selon les ordinateurs de sorte que les utilisateurs doivent se familiariser avec un nouveau langage à chaque fois qu'ils interrogent un nouveau serveur. Pour remédier à cet inconvénient qui découragent certains utilisateurs, une étude menée sous l'égide de la Commission des Communautés Européennes a défini un "langage commun de commande" qui permet d'interroger différents serveurs en employant les mêmes codes d'interrogation. Les serveurs se sont engagés à adopter ce langage commun et plusieurs d'entre eux l'offrent à la place ou parallèlement à leur propre langage.

c) Le service automatisé de renseignements est un autre aspect des services Euronet-DIANE. Il fournit des informations régulièrement mises à jour sur le fonctionnement du réseau, les serveurs et les bases et banques de données; ces informations peuvent être obtenues par terminal, via EURONET, en tapant sur le clavier le numéro d'accès (NUA): 0234307813DN...accompagné du numéro d'identification de l'utilisateur; l'appel se fait donc comme pour n'importe quel serveur.Par rapport aux feuilles de nouvelles publiées par la Commission des Communautés Européennes, ce service présente l'avantage de pouvoir fournir des informations réactualisées quotidiennement si cela est nécessaire. C'est un service totalement gratuit pour l'utilisateur et qui existe dans les six langues de la Communauté.
Développer des services communs dans le cadre multinational et multilingue de la Communauté est l'un des rôles dévolus à la Commission Européenne; aux projets cités ci-dessus, on peut ajouter:
- la sensibilisation des utilisateurs aux avantages de la recherche documentaire en ligne,
- l'encouragement à la création et à la diffusion d'informations,
- la traduction automatique (par ordinateur) des documents,
- des manuels d'utilisation et des aides à la formation,
- les outils multilingues, (vocabulaires, thesaurus),
- la commande par terminal des documents originaux et l'amélioration des services de livraison de ces documents,
- des conditions de ventes normalisées.

III. Devenir utilisateur d'Euronet-DIANE.

On constate qu'avec les nouveaux réseaux il n'est plus nécessaire d'appartenir à une importante société commerciale ou à un "riche" laboratoire pour accéder à l'information. Euronet-DIANE intéresse, bien évidemment et, en premier lieu, les chercheurs qui travaillent en laboratoire, les professeurs, les étudiants, les ingénieurs de recherche; il intéresse également et, au fur et à mesure de sa diversification, il intéressera de plus en plus de membres d'autres professions: responsables d'entreprises, économistes, juristes, journalistes, milieu bancaire...L'intérêt sera d'autant plus grand que les modalités techniques d'accès seront simplifiées. Ci-dessous, nous indiquons les différentes conditions à remplir pour devenir utilisateur d'Euronet-DIANE

a) Matériel

Le matériel nécessaire comporte trois éléments:

- un terminal "compatible télétype" (TTY) avec écran ou/et imprimante: Euronet a été spécialement conçu pour supporter le plus grand nombre possible de terminaux et la catégorie TTY est la plus répandue;
- un appareil téléphonique normal relié au réseau commuté;
- un modem (modulateur-démodulateur) ou un coupleur acoustique qui sera intercalé entre le terminal et l'appareil téléphonique.

b) Procédure

- L'utilisateur doit prendre contact avec le service EURONET des PTT de son pays afin d'obtenir un numéro d'abonné ou NUI.
- Il prendra également contact avec le ou les serveurs qui possèdent les fichiers auxquels il s'intéresse; un contrat (généralement gratuit) sera établi qui l'autorisera à interroger ceux de son choix; un mot de passe lui sera attribué par le serveur pour accéder à ces bases.

La facturation des services est effectuée, d'une part, par les PTT, pour l'usage du réseau et le volume du trafic d'autre part, par les serveurs, pour leurs services et l'information transmise.

Pour aider les utilisateurs ou futurs utilisateurs, la CCE a créé l'Equipe de lancement d'Euronet-DIANE. Son rôle est d'orienter les utilisateurs et de fournir des informations sur les fichiers disponibles,sur les serveurs, sur le réseau et de faire la promotion d'Euronet-DIANE en général...

L'Equipe de lancement d'Euronet-DIANE est à la disposition des utilisateurs qui souhaiteraient davantage d'informations:

Adresse:BP. 777
15, avenue de la Faiencerie - Luxembourg.
Tél.: +352 40221 - Telex: 3511 Diane Lu

Conclusion

Le réseau Euronet concerne actuellement onze pays. Des accords sont à l'étude avec la Suède, la Finlande, l'Espagne...afin d'élargir la communauté des bénéficiaires des services Euronet-DIANE; ainsi, les 50 pays du groupe "Africain, Caraïbes et Pacifique" qui ont signé la Convention de Lomé avec la CEE ont fait part de leur souhait d'être, un jour, associés à Euronet.

Ces quelques indications soulignent l'importance des réseaux à l'égard des échanges d'information. Leurs développements et leurs interconnexions futures élargiront encore leur champ d'activité et augmenteront les services qu'ils rendent déjà aux utilisateurs. Bénéficiant d'une technologie en plein développement, on peut penser, raisonnablement pour les prochaines années, accéder facilement à des connaissances jusqu'ici réservées à quelques privilégiés.

Le colloque d'aujourd'hui qui est consacré aux "réseaux de données en astronomie" permet d'estimer l'ampleur des ressources d'information dans ce secteur. D'autre part, votre communauté par son caractère international rend indispensables les échanges rapides des informations.
Je crois qu'Euronet-DIANE par ses caractéristiques répond à vos exigences de spécialistes; je souhaite donc qu'il puisse apporter sa contribution à la satisfaction des besoins ressentis dans ce domaine.

THE U.K. STARLINK COMPUTER NETWORK

David J. Carnochan
Dept. of Physics & Astronomy, University College London

Six VAX-11/780 computers have been set up at different centres of astronomical research throughout the United Kingdom. These computers are solely for the use of astronomers and they are linked together using special telephone lines. They are proving to be a powerful tool, both in the reduction of astronomical images and in the subsequent data analysis.

StarLink is a network of six computers that are 'linked' together using telephone lines in a 'star' pattern; hence the name 'StarLink'. The six locations are: (1) RAL, the Rutherford & Appleton Laboratories, (2) Cambridge University, (3) ROE, the Royal Observatory, Edinburgh, (4) Manchester University, (5) RGO, the Royal Greenwich Observatory, and (6) UCL, University College London. The computer at RAL is at the centre and the other sites form a five-pointed star.

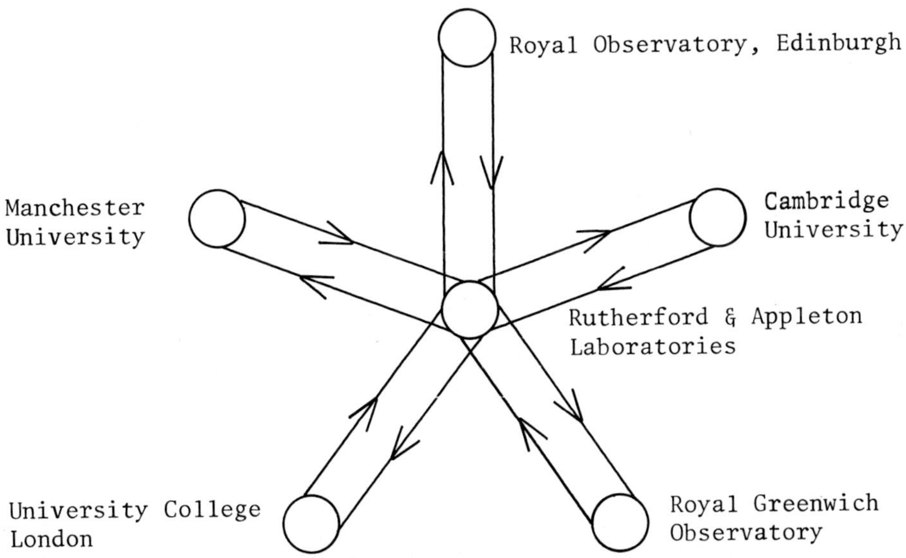

C. Jaschek and W. Heintz (eds.), Automated Data Retrieval in Astronomy, 63–65.
Copyright © 1982 by D. Reidel Publishing Company.

The transfer of data between sites has to go through the RAL node. Thus RAL can communicate directly with the other five computers, and vice-versa. But ROE can only communicate with RGO by following the path ROE - RAL - RGO. The system used to link the computers together is DECnet which is a product of Digital Equipment Corporation (DEC) who make the VAX computers.

Simple linking operations whereby one site serves as a remote terminal are straightforward. An example of this is provided by a Manchester astronomer accessing the RGO computer (via RAL) and using an RGO program to draw finding-charts. However, more sophisticated applications are also possible. An astronomer at ROE can execute a UCL program for analysing IUE spectra, he can take an IUE image held at Cambridge and superpose over its spectrum a selection of spectral lines taken from a line list at RGO. DECnet is a relatively slow network and its use is only encouraged for the transfer of small amounts of data. A spectrum with 1000 data points is transmitted in a few seconds, but a 500 x 500 image takes many minutes.

Many modern astronomical experiments produce two-dimensional images that contain a very large number of data elements. These include photographic plates (which after scanning on a microdensitometer typically result in images of 100,000 x 100,000 pixels); IPCS, the Image Photon Counting System (which typically has images of 1,000 x 1,000 pixels) and, IUE, the International Ultraviolet Explorer satellite (where every image is 768 x 768 pixels). In order to handle these images, each StarLink site is provided with graphics terminals where the data can be displayed & manipulated, both in black & white and in colour. 'Hard-copy' devices are also available so that a permanent record of any image can be made.

Another important consideration of data reduction & analysis is that every image tends to be 'special' demanding to be handled in its own unique way. Consequently, StarLink has been set up to run interactively. Although data can be reduced using 'black-box' programs in batch, most astronomers now prefer to manipulate the data themselves. On StarLink, the results are displayed back on the user's terminal as soon as the data is processed. The astronomer can react to the changed image immediately and adjust the flow of his analysis accordingly. StarLink provides the astronomer with the ability to do the analysis that he wants to do, rather than being limited to what a rigid system provides.

For example, consider the analysis of stellar spectra. For our first image we might be interested in identifying lines and require access to lists of spectral lines. With our second image we might find that the star has emission lines and need to switch over to a program that will fit P-Cygni profiles. For our third star we might be interested in interstellar effects and wish to difference its spectrum with a standard star in order to get at the extinction curve. And with our fourth image we might become a theoretician and not wish to process the image at all, but to 'process' instead a model-atmosphere & line-profile program to get at a best fit model. All of these are possible with StarLink and many have already been implemented and are available now.

Of course, not every piece of analysis needs to use the StarLink network in order to access the other sites. Indeed each site has its own particular interests and the majority of work is carried out as if each computer were a stand-alone machine. RAL controls StarLink, provides much of the hardware & software, and coordinates the different site activities. RAL also serves as the StarLink node for Oxford University where one area of interest is in reducing plates of clusters of galaxies. Edinburgh uses StarLink to analyse photographic plates that have been scanned using the 'Cosmos' microdensitometer, most of these plates are taken with the Anglo-Australian Schmidt telescope. Manchester processes data collected by the Jodrell Bank radio telescopes. RGO works in many areas including the three-dimensional images obtained with a Fabry-Perot interferometer used in conjunction with IPCS, typically for analysing of the velocity distributions in extended objects such as galaxies. UCL is primarily concerned with IUE images and with the IPCS observations of quasars. Cambridge, of course, does a bit of everything.

StarLink is not an isolated network of computers, and connections can be made into the general network that links together the majority of computers within the UK Science Research Council. However the special linking of identical machines is now lost and the StarLink computer then serves as a simple remote terminal.

For the future, the 'Universe' project is researching into the use of 'rings' for communicating large quantities of data between computers. Simple dial-up terminals transfer a few hundred bits per second, DECnet on StarLink can handle a few kilo-bits per second; rings offer the possibility of transmitting mega-bits per second. Rings are used to connect different computers that are physically very close to one another. Packets of data are sent from a machine's station into the ring. The packets pass through the stations of the other computers in a fixed direction around the ring and they leave when they arrive at the requested destination. One ring links the Cambridge University computers and another ring links the UCL computers. One of the stations in the ring can be a ground-station in which case the packet of data can be transmitted up to a satellite and back down into a different computer ring, perhaps half-way around the world. StarLink is involved in the Universe project and the UCL Vax will be used to help investigate the problems of moving bulk data between computers.

ASTRONET. THE NETWORK FOR ANALYSIS AND RETRIEVAL OF ASTRONOMICAL DATA IN ITALY

G. Sedmak
Osservatorio Astronomico di Trieste
Via G.B. Tiepolo 11
34131 Trieste - Italia

ABSTRACT

ASTRONET, the network for analysis and retrieval of astronomical data planned in Italy, is described with regard to the organization of the network, the hardware, and the standard software environment. ASTRONET starts with four VAX centers operating within 1981, to be expanded to six within 1983.

The network includes colour graphics and PDS plates digitization facilities. The software and graphics standards will be set as compatible as possible to major existing astronomical standards. ASTRONET is planned to be operative within 1983 as an open international astronomical facility.

1. INTRODUCTION

The increasing rate of large volumes of astronomical data from remote sources, like astronomical satellites and international ground-based facilities, generated a rapid growth of the number of data processing centers dedicated to astronomy as well as of astronomers pushing for access to further resources.

The analysis of the astronomical data processing scenario in Italy based upon the information contained in the plan of development for the years 1981-82-83 provided by the Group on Astronomy of Italy's Research Council, shows that about 70 % of the resources for national research programs were planned to implement astronomical data processing facilities.

After this analysis, the scientific council of the Group on Astronomy approved a development plan for a national-wide data analysis and retrieval facility for astronomy. Taking into account the geographical distribution of the astronomical research centers and the data processing resources already available, the recommendations of the commission

C. Jaschek and W. Heintz (eds.), Automated Data Retrieval in Astronomy, 67–71.
Copyright © 1982 by D. Reidel Publishing Company.

of Italy's Research Council for the participation of Italian astronomers in the Space Telescope program, and following the impact of UK STARLINK, the structure of this facility was established with a network of hardware compatible systems linking together selected astronomical centers. This network is primarily intended to integrate astronomers into a common standard software environment provided with data base facilities. The distributed processing of substantial amounts of data is not presently considered even if technically possible.

The name of this network is ASTRONET.

2. ASTRONET

ASTRONET is the network for analysis and retrieval of astronomical data planned in Italy to link together into a standard software environment most existing astronomical data processing centers plus some new ones. ASTRONET is also intended to enhance the circulation of information and know-how relevant to astronomy and the integration of multi-center research activities.

The ASTRONET program was started in 1981. The organization of the network is shown in Fig. 1.

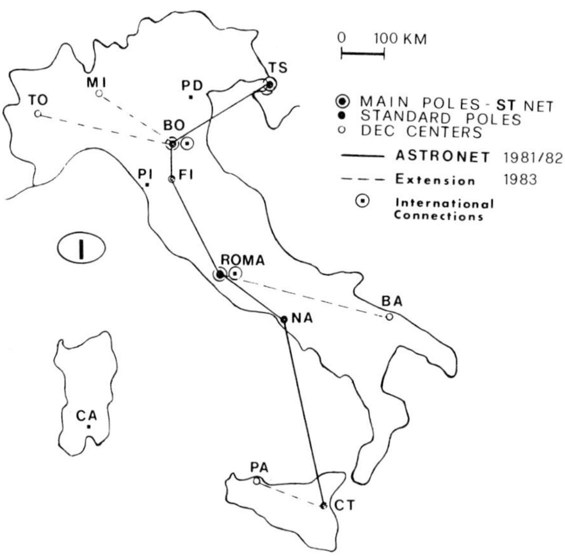

Fig. 1 ASTRONET architecture

The mainframe consists of the astronomical data processing centers of Bologna, Firenze, Roma and Trieste, already equipped with DEC VAX systems, plus the centers of Catania and Napoli, presently equipped with DEC PDP11 computers but to be upgrated into VAX systems within the ASTRONET development plan. Each pole should integrate all the astronomical data processing centers operating in its area. Multi-center poles are foreseen in Bologna and Roma. The mainframe will include special graphics and plates digitization facilities in selected poles. Two AYDIN 5000-series colour graphic sub-systems are already integrated in the Bologna and Trieste poles and two PDS 1010 A microdensitometers in the Napoli and Trieste poles.

Some centers relevant to astronomical research that are not directly included in the ASTRONET mainframe could be connected in the future under an agreement of the parts involved, as well as centers dedicated to research in computer data processing relevant to astronomy.

The network will be operated first by standard DECNET software through switched telephone lines at low data rate. This approach should allow to set up the network and start the standardization of user access to available resources with a minimum expense of time and efforts. In the future ASTRONET is planned to be operated through the commercial network in packet-switching mode depending on the availability of this facility in Italy.

International connections of ASTRONET to other astronomical networks and/or data centers are planned to be implemented through the major poles, including high data rate satellite connections.

3. CONFIGURATION OF POLES

The configuration of the poles of ASTRONET was defined so as to allow effective processing of the large arrays of data corresponding to rapid photometry, high resolution spectrometry, and two-dimensional processing of astronomical images, with a typical array size figure of 512 KB.

The configuration is based on the DEC VAX 32-bit computer under VMS operating system. Two levels of throughput are provided using respectively the VAX 11/780 and VAX 11/750 CPU. Some poles include specific hardware for special graphics and plates digitization. Array processors could be included if proven necessary.

The typical configuration of a full-service pole is shown in Fig. 2 The typical configuration supports a standard DEC VS 11 colour terminal for low-end graphics at 512 x 512 x 3 format. Advanced graphics sub-systems could be included according to the activity planned for the concerned pole. The graphics hardware presently available consists of 4000-series TEKTRONIX storage-type terminals, AYDIN 5000-series colour display generators of 512 x 512 format, and PRINTRONIX and/or VERSATEC printer/plotter hard-copy units.

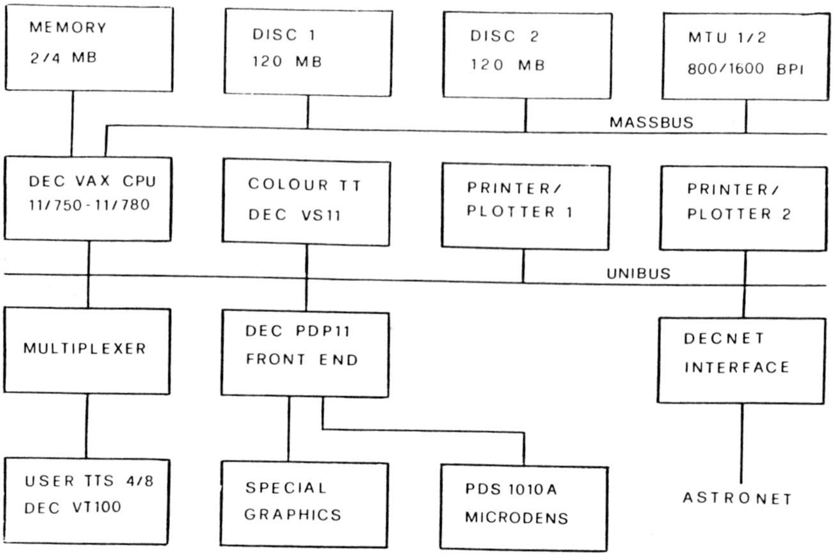

Fig. 2 Configuration of the typical ASTRONET pole with special graphics and plates digitization facilities.

This hardware will be implemented/replaced by state-of-the-art components within the ASTRONET development plan taking into account the international trends and the working experience on astronomical applications.

The plate digitization is made by astronomy-standard Perkin Elmer PDS 1010 A microdensitometers. Graphic tablets and/or TV digitizers will be connected successively for specific applications.

The discs and tapes shown in Fig. 2 are the low-end standard. Large 512 MB discs and 6250 BPI tapes will be included in the configuration within the ASTRONET development plan.

4. SOFTWARE STANDARDS AND GRAPHICS

One main goal of ASTRONET is to accomodate astronomical users into a specific standard software environment for effective access to available resources and integration of contributed software. This environment will be made as compatible as possible to existing major astronomical packages.

The software environment will be organized so as to interface the dedicated application modules to (a) user, (b) graphics, (c) data storage, and, (d) documentation, through standard software interfaces. A specific interactive control language could be implemented within the user interface if a suitable standard will be identified for this purpose. Standard data structures will be in any case identified and imposed to organize the data storage for processing purposes.

The implementation of a distributing data base compatible with and possibly connected to existing major astronomical data bases is planned within the ASTRONET development.

The standardization of graphics will be implemented through a common basic graphics software running on ASTRONET standard graphic terminal. The basic software will be selected taking into account international astronomical trends. A specific high-level graphic package is planned to be introduced and integrated as one main part of the ASTRONET program.

The documentation on the software interfaces standards and the application modules and/or packages available will be maintained in ASTRONET together with the planned data base for on-line access by the users.

5. CONCLUDING REMARKS

The ASTRONET program should be carried out within 1983. This program should be parallel to the development and/or integration of specific astronomical software to be run on ASTRONET.

The coordination of the network will be in charge of a management group composed of one responsible/pole plus representatives of the users. The management will be responsible to the scientific council of the Group of Astronomy of Italy's Research Council, which will be represented in the management group.

The development of ASTRONET will be highly integrated to international astronomical data processing programs. Consequently, recommendations and/or proposals are welcomed.

After successful completion ASTRONET will be operated as an international facility open to all astronomers.

ACKNOWLEDGEMENTS

ASTRONET is based upon the contributions of several persons and institutions, acknowledged here as a whole.

I present ASTRONET in my quality of chairman of the commission for astronomical technologies of the Group of Astronomy of Italy's Research Council.

THE INTERNATIONAL URSIGRAM AND WORLD DAYS SERVICE

Simon P.A
Chairman - Observatoire de Meudon

SUMMARY

Broad outlines of a permanent service involved in the prompt dissemenition of solar and geophysical data.

The Solar Activity has an effect on several modern technics. For instance, during a transient increase of solar UV flux, the trajectory of a low perigee satellite can slightly change, a ionospheric communication link can fade out or some scientific program can only be carried out during some outburst of the solar activity.

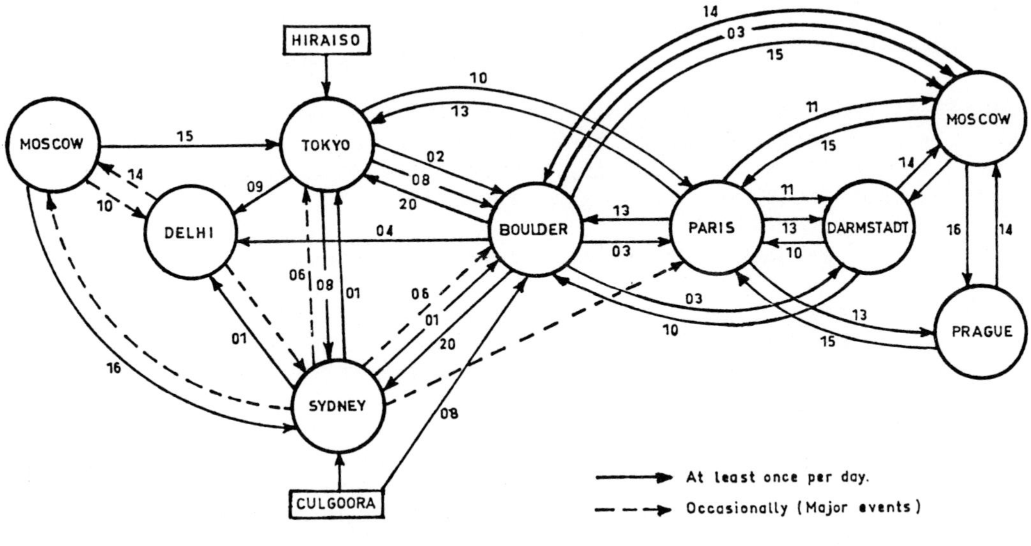

Fig. 1 Daily Schedule of interchange messages between RWCs

C. Jaschek and W. Heintz (eds.), Automated Data Retrieval in Astronomy, 73–75.
Copyright © 1982 by D. Reidel Publishing Company.

As early as on 1928, solar astronomers and geophysicists began to daily broadcast series of event reports from the Eiffel Tower station, in Paris. Presently, around the world, several hundred observatories and stations supply with their reports nine Regional Warning Centers (RWC) which cooperate together under the patronage of the International Ursigram and World Days Service (IUWDS), an organization cosponsored by the three relevant scientifics unions : IURS, IGGU and IAU. The Boulder RWC, selected as a World Warning Agency, is in charge of issuing a daily GEOALERT message established according to the advices of all the RWCs and distributed through the World Meteorological Organization network. In their regional area, the RWCs collect data and reports, issue daily Ursigram messages including data, reports and forecasts which are broadcasted and distributed by telex and telephone to permanent and temporary users and to other RWCs (fig. 1);

```
USIDS 31510 10630 40623 00628 00655 50621 00628
00650 70622 00628 00635
UABSE 36801 01101 30110 90120 01/01

URALN 30509 10701 00169 12010
USSPS 31405 30066 21322 56013 46712 54505 60001 33905
01204 64008 41019 34410 65002 37219 21302 66106 21314
57657 67011 12619 2/901 68001 33412 01205 69005 20106
24409 70001 10419 01202 71001 23115 01205 72001 23604
01203
```

Fig. 2 Samples of encoded messages. USIDS and URALIN report respectively a solar UV burst and an auroral disturbance ; URALIN and USSPS describe radio and optical observations of the solar disc.

Nearly 25 codes are used either to describe data or to report events : after a five letter block of code identification, a series of five digit blocks shows the reporting station, the date of the report and the relevant pieces of information (see fig. 2 above).

The special features of this Service are as follows :

1. An international service working thank to the free cooperation of scientific observers and of Regional Centres established and supported by national agenicies (Telecommunication, Scientific Institutes, etc..)

2. Several fields of sciences are involved : solar astronomy, interplanetary research, physics of Earth environment and geophysics. The main interests are in both the monitoring of the relevant slowly varying components (solar active centres, critical frequencies of ionospheric layers, etc...) and the report of transient events (solar flares, geomagnetic storms, ionospheric disturbances, etc...).

3. Data and reports are distributed promptly, sometimes in real time : they are used by the RWC staffs to issue warnings and to prepare forecasts. These forecasts are related to events which are supposed to occur after a few hours or days according to the category of event : after these delays, the relevant forecasts are obsolete. Likewise, as most of the event reports are related to "events in progress", they cannot be stored by World Data Centers which request reports established after a careful analysis of completed events.

4. We would like to put the final emphasis on the large variety of users : oil companies, telecommunication services, space agencies and technicians or scientists involved in rocket, balloon and satellite experiments, solar astronomers and geophysicists inteterested in world wide programs of coordinated observations, water supply to developping countries, etc...

N.B. List of IUWDS RWCs : Boulder (USA), Paris (France), Darmstadt (FRG), Praha (Cseckoslovakia), Warsow (Poland), Moscow (USRR), New Dehli (India), Tokyo (Japan) and Sydney (Australia).

C. NEW HARDWARE

NEW DEVELOPMENTS IN DATA STORAGE.

P. J. Grosbøl
European Southern Observatory
Karl-Schwarzschildstr. 2, D-8046 Garching, FRG.

ABSTRACT

It is estimated that up to 100 Gbytes of primary data from digital detectors have to be stored each year. The amount of reduced data is at least one order of magnitude smaller. Although the storage density for magnetic recording can be made higher, only the optical technic can provide a substantial denser medium than the photographic emulsion. It seems likely that optical read-only devices will be developed for archival storage of data in this decade. Magnetic recording of data will still be preferred whenever changing of the information is important.

1. INTRODUCTION

The largest source of astronomical data has until now been photographic plates. This seems also to be the case in the next decade although several new detector systems have been developed recently. These detectors have a much higher quantum efficiency than a photographic emulsion , however , their total data acquisition rate is at least one order of magnitude smaller than that of a Schmidt plate due to the size difference. Whereas the photographic emulsion also has been very good for data storage, new technics in magnetic and optical recording of digital data have made it possible to achieve significantly higher bit densities and smaller access times. Although there is a strong development in data storage for small systems this paper will mainly deal with the new technics for storage of large amount of data.

2. STORAGE REQUIREMENTS

To evaluate modern data storage technics the storage requirements in astronomy must be estimated. In Table 1 the amount and rate of data for different detectors are given. These values may vary considerably

C. Jaschek and W. Heintz (eds.), Automated Data Retrieval in Astronomy, 79–83.
Copyright © 1982 by D. Reidel Publishing Company.

depending on the type of observations being done.

Table 1 : Data rates in astronomy

Source	Mbytes/frame	frame/night	Mbytes/night
Schmidt plates	1000	5	5000
9 cm McMullan E. Camera	100	20	2000
Prime focus plates	500	10	5000
CCD camera	1	50	50
TAURUS	5	20	100

The total amount of data per year is obtained by multiplying these values with the number of telescopes and usable nights (i.e. between 100 and 1000). The largest single source of data in astronomy is photographic plates. Since they already provide the archival storage themselves the immediate problem is storage of between 10 and 100 Gbytes per year from digital detectors (i.e. CCD cameras). An equivalent amount of data is expected from the Space Telescope. The main problem is primary data since after final reduction they become orders of magnitude smaller.

3. COST, SPEED AND CAPACITY

There is a strong correlation between cost, access time and capacity of storage devices which can be seen in Figure 1. When the bit density is made higher the cost and access time will normally decrease while the total capacity of the device becomes larger. Therefore, it is interesting to look on the development of bit density as a function of production year as seen in Figure 2 (Rodriguez 1981).

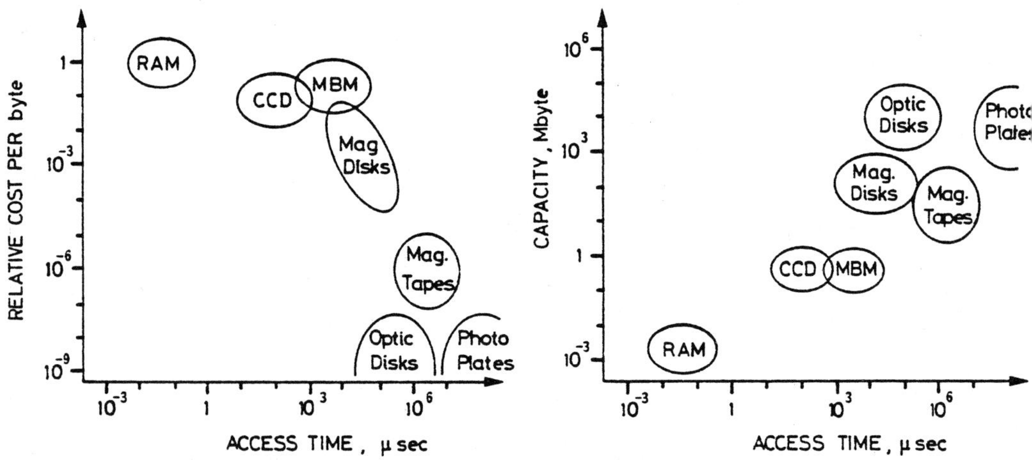

Figure 1 : Cost and capacity of storage as function of access time.

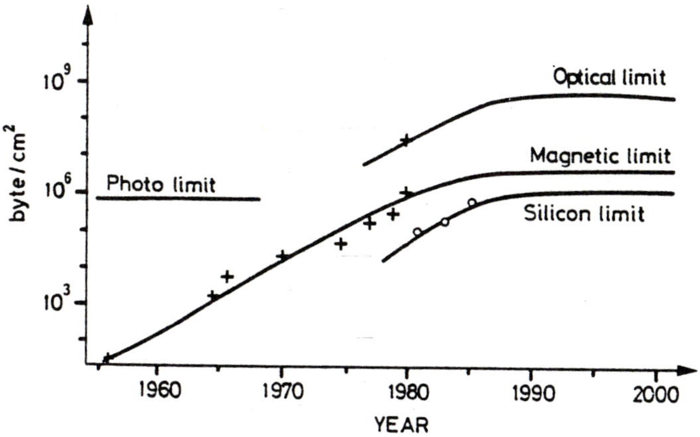

Figure 2 : Bit density as function of production year.

Only the limit for photographic emulsions is reached while the other storage media still can be improved. It seems likely that optical disks will be the preferred media for archival purposes (i.e. read only applications) when they are fully developed.

4. SILICON TECHNOLOGY

Two new kinds of memories based on semiconductors have been introduced, namely : Charge-Coupled Devices (CCD) and Magnetic Bubble Memories (MBM). Their performance (i.e. access time and storage capacity) places them between Random Access Memories (RAM) and disks. The main difference is that MBMs are nonvolatile because the information is stored as small magnetic domains (see, Snigier 1979 and Haggard 1981) while CCDs are volatile (Bhandarkar et al. 1979). Since their maximum storage capacity is about 1 Mbyte they are mainly used as buffers between the fast memory and on-line mass storage devices.

5. MAGNETIC TECHNOLOGY

The demand for higher densities of magnetic recording of data has created new types of disk and tape drives. Both the size of the recording head and its distance from the magnitic surface have to be decreased in order to achieve higher densities. The technology of thin film heads has reduced their thickness to less than 100 microns. The decrease of the distance between head and recording surface to the order of 50 microns has ,however, made it necessary to protect disk surface from contamination. This development has given rise to the fixed 'Winchester' type disks (see, Keith Plant 1980 and Shershow 1980). Disk drives with a capacity above 500 Mbytes are already available using this technic.

The fixed disks have created a need for fast, high density tape drives to be used as back-up devices. Although the recording density can be increased the standard inter-block gaps on 1/2" tapes make it less efficient. The new streaming tape drives which record the data continuously (Valiant 1979) have solved this problem while some of the compatibility has been lost. These streaming drivers can offer capacities over 100 Mbytes on a 600 feet 1/4" tape cartridge and data rates of several Mbytes per second. These features make them quite attractive for internal use in data centers.

6. OPTICAL TECHNOLOGY

A major part of astronomical data have been stored optically in the form of photographic plates. This analog storage has two advantages compared to present technics, namely : its high information density and its excellent archival quality. The problems, using glass plates as mass storage, are the very long access time and the difficult handling. The new developments in digital optical data storage seem within this decade to be able to provide a storage medium which both has up to 100 times higher bit density, a much shorter access time, and an archival quality. Although these optical methods are not able to change data already written they are of high interest for archiving astronomical data. The two main technics being holography and spot recording (Redderson and Ralston 1980) are discussed below.

The holographic technic records the standard Fourier transform of an array of bits on a photographic film. Different methods have been used to perform the transform. Most recently the Wideband Holographic Recorder/Reproducer system by Harris Corporation has employed an acousto-optic page composer as signal distributer and generated the Fourier transform interferometrically. This system has demonstrated a recording rate of over 100 Mbytes per second while the bit density is that of photographic emulsions. For astromony a high bit density is much more important than a high transfer rate. Thus, it seems less likely that the holographic recording technic will be of interest.

The other optical storage method is spot recording where the data are stored on a disk by altering its surface using a focused laser. In this way densities higher than 100 Mbytes per square cm (i.e. at least 100 times denser than photographic emulsions) can be reached. The technic developed by Philips laboratory and RCA records the data on a master disk with a thin film of a ablative material such as tellurium or bismuth. Copies in plastic can then be made from this master disk. In the case of digital data storage where only one or a few copies are needed the quality of the master is important. Unfortunately, the abletive materials now used show a significant degradation within a few years even if they are protected with overcoating of glass or plastic. A new technic developed by 3M Corp. may, however, provide the archival quality needed for making optical storage of data usable in astronomy. Instead of changing the reflectivity of the disk surface by removing material, small bubbles are made in the top layer consisting of a material with an extremely high melting point (Robbins 1981). The bubbles are formed by focusing a laser on the underlying material which then outgasses and presses the surface out. If these types of optical disks are used an on-line storage of several Gbytes per drive is possible.

7. CONCLUSION

Magnetic tapes and disks will still be the main data storage media in the coming years. The increased density offered by the thin head technology makes it possible to have all frequently used data on on-line disks while tapes are used for transportation and long term storage of data. The 1/2" tapes will remain important due to the large amount of data already on them and to their standard format which makes them portable. Streaming tape drives may ,however, take over most of the internal storage such as disk back-ups. With the increasing amount of data from digital detectors an archival storage medium with very high bit density is needed. It is likely that optical disks can provide this within the present decade. In that case not only primary data from detectors would be stored on such disks. Also some often used photographic data (e.g. the sky surveys) could be transferred to digital form to decrease the access time for them.

8. ACKNOWLEDGEMENTS

I would like to thank F.Middelburg for many interesting discussions during the preparation of this paper.

9. REFERENCES

Bhandarkar,D.P.,Barton,J.B.,and Tasch,A.F.Jr.: 1979, IEEE Computer Vol.28 no.1,pp.16-23.
Ford,R.: 1980, Digital Design Vol.10 no.5,pp.46-51.
Haggard,R.: 1981, Digital Design Vol.11 no.2,pp.31-32.
Hoagland,A.S.: 1979, IEEE Computer Vol.28 no.5,pp.12-18.
Keith Plant,A.: 1980, Digital Design Vol.10 no.5,pp.62-68.
Redderson,B.R. and Ralston,L.H.: 1980 Optical Engineering Vol.19, pp.199-204.
Robbins,W.: 1981, Laser Focus Vol.17 no.6,p.10.
Rodriguez,J.: 1981, Electronic Design Vol.29 no.13,p.36.
Shershow,H.: 1980, Digital Design Vol.10 no.5,pp.52-60.
Snigier,P.: 1979, Digital Design Vol.9 no.12,pp.88-91.
Valiant,W.: 1979, Digital Design Vol.9 no.12,pp.94-96.

D. RECENT SOFTWARE DEVELOPMENTS

SYSTEM SOFTWARE APPROACHES TO THE ANALYSIS OF MULTIDIMENSIONAL DATA STRUCTURES

Rudolf Albrecht
Institute for Astronomy, University of Vienna
Space Telescope Science Institute, Baltimore

ABSTRACT

In this review paper on software techniques, the concepts of "software system" and "data base interface" are defined in the context of astronomical data analysis requirements. Principles of software design and maintainance over are discussed.

A deliberate effort was made to stay as close as possible to the astronomical application and not deviate too much into computer science.

1. INTRODUCTION

The analysis of multidimensional data structure is more commonly known as Image Processing. Image processing is of growing importance for modern astronomical data analysis. At the Space Telescope Science Institute, a data analysis system for data generated by the Science Instruments of the Space Telescope is being defined. Although I will in this presentation frequently refer to Space Telescope oriented applications, this is meant as an example only; the development has just begun, and the design has not been finalized (Albrecht, 1981)

2. DATA

Image-type data usually come from panoramic detectors. These are devices which transfer an intensity distribution into an array of digitally represented data points. Already, such detectors are in widespread use and they will become more common with evolving technology and lower prices.

But even data obtained with the more conventional technique of the photographic process quite often are available in digital form. Microdensitometers are being used to digitize astronomical plates. Aside from the fact that data generated by two-dimensional electronic detectors are linear with intensity and data obtained by digitizing a

C. Jaschek and W. Heintz (eds.), Automated Data Retrieval in Astronomy, 87–97.
Copyright © 1982 by D. Reidel Publishing Company.

photographic plate are recorded in density units, there is no difference in principle between processing data from two-dimensional detectors and processing microdensitometer data.

Data in such sets consist of different elements. The data points of the image itself are usually referred to as pixels or picture elements. In addition to the pixel data, there is additional information needed to interpret the pixels. This part of the data set is known as the header. Note that there is no difference between pixel data and header data as far as their presence in the data set is concerned.

Other data logically belonging to the header data are the engineering data. Engineering data are not of immediate astronomic interest but rather refer to events on, for example, a spacecraft. The engineering data might become astronomically relevant once they indicate situations which affect the image data. A recent example is the temperature of the onboard computer of the IUE satellite.

There are auxiliary data, for instance, comments given by the observer at the time of measurement. Auxiliary data can also be accumulated during the processing of the data. In some systems such auxiliary data are collected in a trailer file. Ideally, the trailer file is open-ended, so all additional information can be accommodated.

The pixels within an image are usually structured in some approximation of a rectangular coordinate system. In theory, it could be a system other than a rectangular one, but for technical reasons, this is the system most often used. Within such a rectangular coordinate system the pixels are arranged in lines, lines are grouped to frames, and sometimes frames are grouped to cubes. There is no theoretical limit to the dimension of a data set, but for practical reasons, data dimensionality ends usually at three.

Examples for data cubes are: several two-dimensional frames, stacked in time. Another case, occuring very often in radio-astronomy, is a number of two-dimensional frames, stacked in frequency, to examine velocity distributions in a field.

Data coming from the detector or a microdensitometer are usually represented in integer notation. For further data analysis it is useful to convert the data into floating point notation. Although this is somewhat more costly in terms of storage and processing speed, it more than pays for itself in terms of ease of programming and safety during the reduction. Anybody who has ever divided a heavily exposed flat field frame into a low signal data frame will appreciate that.

A serious problem in image processing is the quantity of the data. A typical frame of 500 x 500 pixels holds a quarter million data points. As we begin to arrange many such frames into cubes, the data quantity grows by orders of magnitude. It is obvious that efficient strategies have to be used for retrieving such data from a storage medium.

Most data are stored on magnetic tape as the result of a data acquisition process. For data analysis, they have to be transferred to a faster medium, which today is the magnetic disk. On the disk, the data have to be arranged in a structure convenient for retrieving them and in a format suitable for analysis. Note that there are two different data formats: an internal format and an external format. The internal format is the representation of the data on the system bulk storage. Different strategies can be used here depending on system design. Header files can be connected with the image file or kept separate. Auxiliary files can be kept in image format or in the operating system provided format. The disk can be structured in block format or data can be arranged in some kind of hierarchy.

Completely independent of the internal format is the data representation on an external storage medium, for instance, a magnetic tape for transporting images from one system to another. An agreed upon and already widely used standard is the FITS (Flexible Image Transport System) format (Wells and Greisen, 1979). For anybody transporting images, I strongly recommend adhering to the FITS standard. Not because it is an extremely elegant or philosophically pleasing design, but because it works and it is already supported by several major institutions.

3. DATA ANALYSIS REQUIREMENTS

A data analysis system, such as the one designed to reduce the data produced by the Science Instruments of the Space Telescope, must be designed in such a way as to, in principle, allow the full exploitation of the data. This is not to say that at the very beginning the system has to be capable of analyzing all data in all possible ways, but it means that the system design has to allow for later upgrading of the existing software. Such improvements are necessitated both by the fact that we learn more about how to treat the data as we gain experience with the new instruments, as well as by developments in the field. It is therefore imperative that the design of the system does not preclude later updates of the baseline.

There also has to be a healthy margin of capacity. If the system resources are only sized to handle the requirements of the baseline system, very soon user demands will outgrow the hardware capabilities of the system. This usually results in efforts to circumvent hardware limitations by software. Not only does this make the system much more complicated and much harder to maintain, in the long run, it will also be more expensive in terms of effort and manpower. Today, hardware has become very affordable and no attempt should be made to economize on hardware.

One of the major requirements of any data analysis system is that it must be user-oriented and problem-oriented. Especially at visitor-oriented institutions, the scientist cannot be expected to familiarize himself with rules dictated by and tailored to a particular computer system within the short time he spends at the observatory. In other words, the command language must be as close as possible to plain English, and the individual commands must be meaningful in the context of astronomy.

The data analysis system also has to be interactive. This means that it must be possible for the astronomer to explicitly initiate every step in his reduction procedure and check the result in near real time. Depending on these intermediate results, the astronomer will decide on and initiate execution of the next step. This procedure makes it possible for the astronomer to determine a reduction procedure approprate for his data.

Of course, interactivity also has disadvantages, especially when dealing with image data: given the amount of data points, "near real time" can be up to several minutes, depending on the operations performed. Interactivity becomes dangerous as soon as machine reaction time exceeds the human attention span, because the user will engage in other parellel activities, ultimately losing control over the reduction procedure.

Another disadvantage of interactivity is that, by definition, it requires the physical presence of the user during the reduction procedure. While this is desirable during establishing a proper procedure, it will slow down routine data reduction to an unacceptably low throughput rate. To remedy that, it must be possible, after interactively establishing the best reduction procedure, to switch to a mode of processing, in which the system executes (loops through) a set of reduction procedures for a large number of data frames. This capability is called batch-processing, or macro-processing.

A key issue in every data analysis system is the system documentation.

Not too long ago, the typical scientific program did not even have comments within the program, and there certainly was not any kind of high-level documentation. This situation has somewhat improved, but it is still far from ideal.

In addition to the documented source code, there have to be at least three different manuals: The User's Manual, which is designed to tell the user (the astronomer) how to operate the system, and, in particular, exactly what kind of operation is performed on his data; there has to be an Application Programmer's Manual, explaining how to write, maintain, and modify application programs for the system, how the data are stored on disk, and how to invoke system services; and finally, a System Programmer's Manual is required, explaining matters like the system architecture.

It is imperative that the documentation be maintained along with the corresponding programs, so no gap will develop between the software and its documentation. For the documentation to keep pace with rapidly developing software, this means that word-processing and self-documentation techniques have to be employed.

Another form of documentation which has to be available in an interactive data analysis system is on-line documentation. This form of documentation is usually referred to as the "Help" function. Such a system feature allows the user to ask for help and explanations from the system at any time during the analysis session. A well-designed help function, guiding the user through the system, eliminates the need to consult the printed version of the User's Manual to a high degree and can make the data analysis much more efficient.

Information also has to be provided in a different area: on the scientific background of the system operations. Not only is it important to know how to operate a program to determine the instrumental magnitudes of stars in a two-dimensional frame, for the scientist it is even more important to know which method (algorithm) is used to determine these magnitudes. In other words, application programs of a data analysis system should not be used as black boxes. Of course, this is only possible to a certain degree--it is, after all, not possible to understand just what every single statement in every single program does. However, sufficient information about the program has to be provided to change the "black box" at least to a "gray box." Only in this way will it be possible for the scientist to assume responsibility for the scientific results derived from his data.

4. SYSTEM SOFTWARE

In the case of a computerized data processing system, there are usually two levels of system software to consider. The first level is the host operating system. This operating system is usually vendor-supplied and consists of an ensemble of programs which control and supervise the computer. This host operating system maintains a directory of files on the disk, interprets and executes user commands and provides a number of system services. It is called the host operating system because it provides the environment in which the actual data processing system works. While the host operating system is computer oriented and the commands it accepts are tailored very much to the computer system, the actual data processing system software is much more oriented towards the actual application. The structures of both the host operating system and the data analysis system are very similar. Both consist of a central program, which understands and interprets commands submitted for execution by the user. These commands are executed by invoking other programs known as application programs (Fig 1). The application programs are loaded into the central memory from the disk and exit after execution. Control is then transferred back to the command decoder (Fig 1) There are two reasons for having a data analysis subsystem residing within a host operating system environment: one reason is that the commands of the data analysis subsystem can be made much more understandable to the user, the astronomer; the other reason is that in this way the data analysis software subsystem can be made transportable between computers using different host operating systems.

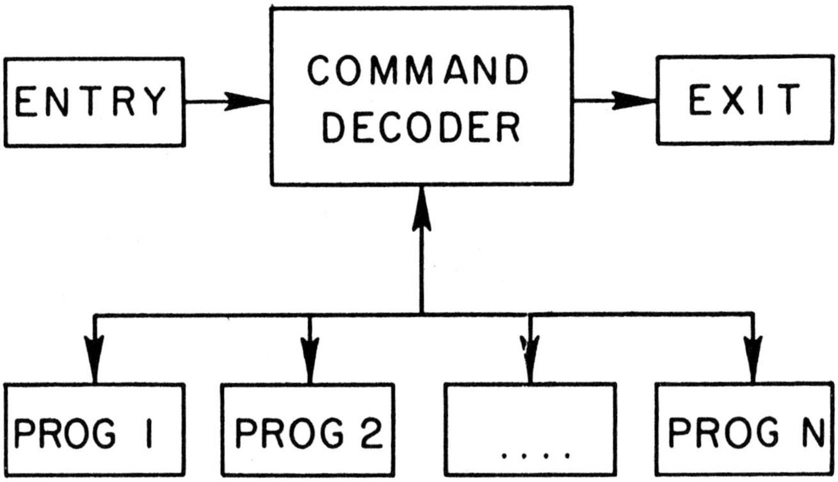

FIG. 1: Overall System Structure

A similar consideration applies to the application programs. As indicated in Figure 2, the central part of an application program consists of the algorithm, for example, a clever way of obtaining a stellar magnitude in crowded fields. This algorithm needs connections to the outside world. User parameters, for instance stellar coordinates, have to be transferred to the algorithm. Access to the data base where the pixels are stored has to be provided for the algorithm. The algorithm has to have a library of mathematical and astronomical routines available for the program. Finally, an interface must exist between the algorithm and the host operating system. This interface provides the algorithm with access to system services of various kinds, for instance a system time service, but also to more complicated system services like disk I/O. This modular structure of the application programs ensures transportability of such programs between different data analysis systems. It is also helpful during the development of the programs. Modular application programs with minimum

on all computers without modifications. However, following a few rather basic rules, it is easily possible to develop software that will run on many computers with only small modifications. Following those basic rules also has another beneficial byproduct: it forces a modular, structured design, which makes the software easier to develop, to understand, and to maintain.

Adequate software maintainance is a key issue, especially for developing software in a research environment, where input comes from many different sources, and the problems to be solved themselves are changing. A program, which is, for instance, developed at the Wide Field Camera Instrument Team, even though it may be transportable, is not necessarily usable for Faint Object Camera Data: the WF/PC generates 12-bit data, while the FOC generates 16-bit data. Even after the program has been generalized to handle data from both instruments, it will have to be changed again to follow suggestions of the astronomers, who have been using it.

It is also clear that such changes, even though they might be beneficial and actually improve the program, have to be implemented with considerable care. Evidently, it is impossible to replace a particular program by another program, even a better one, just as it is impossible to replace the photomultiplier in a photometer in the middle of an observing run.

Such changes in the software have to be made in a carefully controlled fashion. First, it has to be made certain that the requested change to the program does not jeopardize its performance otherwise (e.g., increasing processing time by an order of magnitude). After making the change and implementing it, the new version will have to be tested, both to make sure that the change produces the desired results, but also to verify that other results produced by the program have not been impacted by the change. Finally, the program documentation has to be updated and the program is released to the users. This process is called Configuration Control. It is one of the most important aspects of a large software development project.

Applying these considerations to the problem of data base interfaces, we find the following situation: at the device driver level, the interface is machine-oriented and thus difficult to standardize and certainly not transportable. At the highest level, which is visible to the applications program, standardization is an absolute necessity. This leads to the concept of interface routines consisting of several layers of software between the standardized front end and the machine dependent device driver. Using this concept, it is even possible to maintain transportability of the uppermost layers.
Figure 3 shows this concept as used in the Tololo-Vienna Interactive Image Processing System. On the highest level, the user accesses the image like a large FORTRAN array, using an image name and the pixel coordinates as subscripts. This makes it very easy for the astronomer to write programs performing operations on pixel data.

```
PDP-11

FRONT END:      IPIX=IMAGE (IX,IY)
STRING I/O:     CALL QREAD (LINE,IBUF,IWORDS,IERR)
BLOCK I/O:      CALL QIO (.....)
DEVICE I/O:     Not accessible

HARRIS-DATACRAFT

FRONT END:      IPIX=IMAGE (IX,IY)
STRING I/O:     CALL QREAD (LINE,IBUF,IWORDS,IERR)
BLOCK I/O:      CALL IFXXYY (.....)
DEVICE I/O:     Not accessible

VAX, PHYSICAL I/O

FRONT END:      IPIX=IMAGE (IX,IY)
STRING I/O:     CALL QREAD (LINE,IBUF,IWORDS,IERR)
BLOCK I/O:      CALL SYS$QIO (.....)
DEVICE I/O:     Not accessible

VAX, VIRTUAL I/O

FRONT END:      IPIX=IMAGE (IX,IY)
STRING I/O:     } Mapping
BLOCK I/O:      } Hardware
Device I/O:     Not accessible
```

FIG. 3: Layered interfaces for different host computers.

As can be seen, in this particular interface even the second level is standardized, making the complete first layer transportable. Contact with the host operating system is established only at the third level, where system services are being used to access the disk. This is not applicable to virtual I/O on the VAX, where the VAX-specific mapping features are being used in the first level.

The question of the efficiency of such interfaces has been raised. Clearly, the fastest way of accessing the data is to skillfully use the machine-dependent features of the device driver. However, it is also clear that most astronomers are not willing to write programs at this level.

On the other hand, there is undoubtedly some overhead associated with the high level access of the data base, sometimes slowing down processing considerably. The solution we have adopted is to make the lower layers also accessible to the application program. Thus, if it becomes necessary to speed up a program developed by an astronomer, using front end calls, a system programmer can replace them by faster, low level access calls.

6. CONCLUSION

The recent advances in hardware technology and their application to astronomical data acquintion as examplified by panoramic detectors and orbiting observatories, have brought about the necessity to refine the software tools needed to analyze the data. These tools exist, however, astronomers have not quite learned how to use them efficiently. It must be emphasized that the long term return on investments in hardware and manpower ultimately depends on the quality of the software used for the analysis of the data.

References

Albrecht, R. 1981, Proceedings of the Space Telescope Science Data Analysis Reconnaissance Meeting. ST ScI, Baltimore, MD.

Wells, D., Greisen E., 1979, FITS: A Flexible Image Transport System. In: Image Processing in Astronomy, G. Sedmak et. al. (Eds). Osservatorio di Trieste, Italy.

Allen, R.J. and Ekers, R.D. 1980, Creating Durable Software in a Research Group Environment. In: IUE - Data Reduction, Weiss et. al (Eds.) Vienna, Austria.

A FAST METHOD TO RETRIEVE DATA FROM A LARGE STAR CATALOGUE FILE

A. Schuetz
European Space Operations Centre (ESOC)
Darmstadt
West Germany

ABSTRACT

A method optimized for the access from a large catalogue of stars lying in a given cap of the celestial sphere is described. The performances of the method applied to the SAO catalogue are presented.

1. THE CELESTIAL CUBE

A very frequent type of request to a star catalogue is the access to all stars lying within a cap of a given radius and centered on a given point on the celestial sphere. A partition of the sphere is hence desirable. Identical areas can be obtained by central projection of a regular polyedre inscribed in the sphere. Out of the five regular (platonic) polyedres namely the tetra-, hexa-, octa-, dodeca- and icosaedre, the hexaedre (alias cube) is selected because of the trivial trigonometric formulation involved.

The projection of the cube onto the sphere is called the celestial cube. The cube being oriented such that the centres of the faces are aligned with the 3 reference axes, the faces can be named +X, +Y, +Z, -X, -Y, -Z or 1, 2, 3, 4, 5, 6 respectively, see fig. 1.

2. PARTITION OF THE SPHERE INTO ZONES

A further partition of the faces is required in order to obtain zones of size compatible with the area to be most frequently accessed. On each face of the cube a grid is formed from two sets of equi-distant great circles called 'meridians' as depicted in fig. 1 .

The partitioning factor N defines the number of segments on a vertice. Each face is then formed of $N \times N$ zones, the distance between two adjacent 'meridians' being $DA = 90^\circ /N$. It should be noticed that these zones are not identical.

C. Jaschek and W. Heintz (eds.), Automated Data Retrieval in Astronomy, 99–101.
Copyright © 1982 by D. Reidel Publishing Company.

3. SYSTEM OF CO-ORDINATES

It is here convenient to define a direction S by the following parameters:
- face identification (+X, +Y, +Z, -X, -Y, -Z);
- two azimuth angles.

On fig. 1 the direction S may then be defined by:
- face +Y;
- azimuth angle A1 measured from +X towards +Y;
- azimuth angle A2 measured from +Z towards +Y.

The conversion from unit vector (X, Y, Z) to face and azimuth is trivial: The face number is driven by the greatest component in absolute value and by the sign of that component while the azimuths are simply arctangent functions of the components.

4. ZONE IDENTIFICATION

A zone is identified by face number I (1 to 6), line number J (1 to N) and column number K (1 to N) or by a unique number NZ.

$$NZ = ((I - 1) * N + J - 1) * N + K \qquad (1)$$

Given a direction S (face, A2, A2) the zone number is computed by (1) using: I = corresponding face number (1 for +X, 2 for +Y,...)

$$J = (A2 - 45^{\circ})/DA + 1 \qquad (2)$$

$$K = (A1 - 45^{\circ})/DA + 1 \qquad (3)$$

5. ACCESS BY CAP

In order to fix a maximum admissible radius RM for a cap and hence to limit the number of zones covered, the following constraint is defined: For a given partitioning factor N a cap centered on any direction S should cover at most the zone containing S and its eight contiguous zones.

It can be shown on fig. 1 that the worst case is when S is on the middle of a vertice of the cube where:

$$RM = \text{arcsine } (\text{sine } DA \text{ sine } 45^{\circ}) \qquad (4)$$

Under this constraint the identification of zones covered by a cap of direction S and half-angle R involves the following steps:
- identify the zone (I,J,K) containing the direction S (face, A1,A2)
- compute the half-angles DA1, DA2 subtending the cap from +Z and +X resp.

$$DA1 = \text{arcsine } (\text{sine } R/\text{sine } A2) \qquad (5)$$

$$DA2 = \text{arcsine } (\text{sine } R/\text{sine } A1) \qquad (6)$$

- compare the azimuths of the meridians tangent to the cap ($A1 \pm DA1$, $A2 \pm DA2$) with the azimuth of the meridians of the grid surrounding S and retain eventually 'lateral' zones of the form ($I, J \pm 1, K$), ($I,J,K \pm 1$).

- If zones of both forms are retained check also if 'diagonal' zones of the form $(I,J \pm 1, K \pm 1)$ are covered by the cap.

6. ACCESS BY BAND

The access by band or wedge is a generalisation of the access by cap. The band defined by a pole, tilt angle, half-width and azimuth limits is overlapped by a set of caps, the scan step being equal to the radius of the cap, see fig.1 .

The relation between half-width and radius is:

$$\text{cosine}(HW)=\text{cosine}(R)/\text{cosine}(R/2) \tag{7}$$

7. PERFORMANCES

The SAO catalogue (260000 stars) was used to create a file containing for each of the 339 x-ray sources of the 4th UHURU catalogue all stars lying in a cap of 1.5° radius. The total task required on a CII-10070 computer 16 min. of resident and 150 sec. of CPU time.

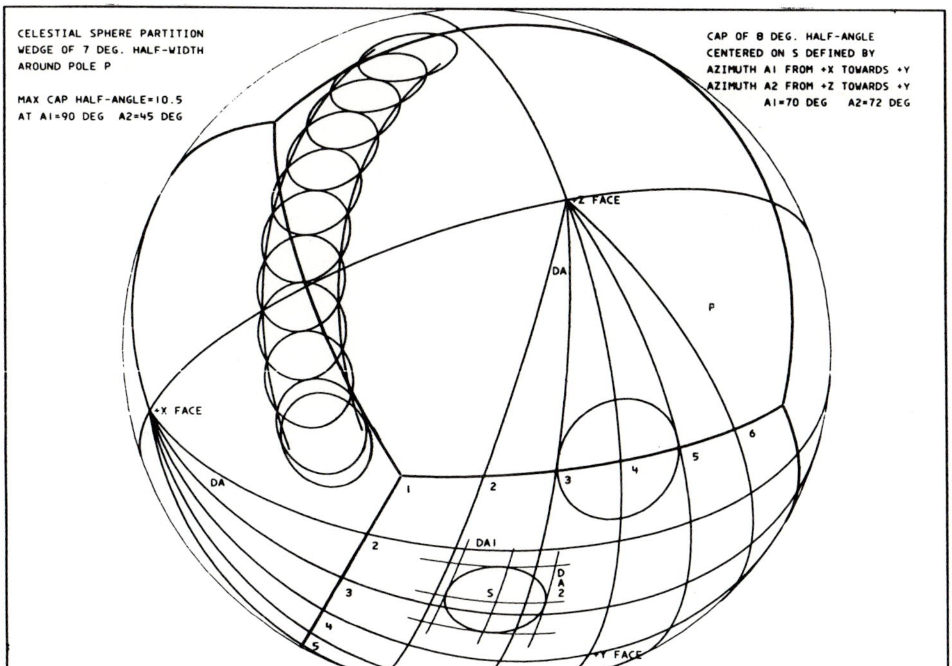

FIG. +Y FACE PARTITION INTO 6×6 ZONES

UNIVERSITY COLLEGE LIBRARY CARDIFF

ON SYSTEMS OF STANDARDS*

Heinrich Eichhorn
Department of Astronomy, University of Florida, Gainesville

Rigorous formulations closely following current practice, are proposed for the absolute and the relative estimation of certain target quantities. A system of standards is defined as a set of estimates that has certain properties. This is followed by a discussion of the meaning of "reducing estimates to a specified system" and certain procedures are suggested for making this process less ambiguous. It is finally shown that relative estimates can be employed for replacing a system which is based on absolute measurements alone by another, more accurate system.

1. ABSOLUTE AND RELATIVE ESTIMATES

There are in principle two ways in which data can be estimated. One of these makes no use of previously existing estimates of like data and leads to *absolute estimates*, the second uses previously existing estimates of like data for calibrating the new data, thus leading to *relative estimates*.

Absolute estimates are obtained in the following way:

Let the *target variable* a have a unique numerical value for each element A of the *object set* $\{A\}$ (if $\{A\}$ were the set of all stars, a might be, e.g., the luminosity.) Consider a set $\{\vec{x}^*\}$ of measurements (that is, estimates which may be regarded as unbiased without restricting generality) of quantities whose true values are $\vec{x}$. Let $\{\vec{p}\}$ be a set of unknown parameters and assume the existence of a model function Φ, that is, a set of functional relationships

$$\Phi(\vec{a}, \vec{x}, \vec{p})=\vec{0} \tag{1}$$

between the vectors $\vec{a}$, $\vec{x}$, $\vec{p}$. Estimates $\vec{a}^*$ are called *absolute* if it is possible to find estimates $\vec{p}^*$ for $\vec{p}$ simultaneously with $\vec{a}^*$ by solving

$$\Phi(\vec{a}, \vec{x}^*, \vec{p})=\vec{0} \tag{1a}$$

C. Jaschek and W. Heintz (eds.), Automated Data Retrieval in Astronomy, 103–108.
Copyright © 1982 by D. Reidel Publishing Company.

for $\vec{a}$ and $\vec{p}$.

Estimates are called *relative* if the pattern for their determination is as follows. Let $\vec{a}_R$ be a vector composed of some components of $\vec{a}$ and let $\vec{a}_R{}^\dagger$ be a previously determined estimate of $\vec{a}_R$. Estimates $\vec{a}*$ of $\vec{a}$ and $\vec{p}*$ of $\vec{p}$ are now obtained from solving the relationships

$$\Psi(\vec{a},\ \vec{a}_R{}^\dagger,\ \vec{x}*,\ \vec{p})=\vec{0} \tag{2}$$

for $\vec{a}$ and $\vec{p}$, where the functions Ψ must be such that the variables represented by $\vec{a}_R{}^\dagger$ cannot be eliminated from them such that $\vec{a}$ and $\vec{p}$ could then still be calculated from the remaining system. (Otherwise, the $\vec{a}*$ resulting from this calculation would be absolute estimates.)

Very often the estimation of $\vec{a}*$ and $\vec{p}*$ can be carried out in steps, as follows. Assuming a relationship of type (1) between $\vec{a}$, $\vec{x}$ and $\vec{p}$, we first find $\vec{p}*$ from solving

$$\Phi(\vec{a}_R{}^\dagger,\ \vec{x}*,\ \vec{p})=\vec{0} \tag{3}$$

for $\vec{p}$, and then calculate $\vec{a}*$ from solving

$$\Phi(\vec{a},\ \vec{x}*,\ \vec{p}*)=\vec{0} \tag{1b}$$

for $\vec{a}$.

2. DEFINITION OF A SYSTEM OF STANDARDS

Consider a subset of $\{A\}$, the *standard set* $\{A_S\}$, such that for all of its elements, a set of estimates $\{\vec{a}_S{}^\dagger\}$ of the target variable is available. These estimates will be called a *System of Standards* (or a *system*) if the following conditions are satisfied. There exist non-empty subsets $\{A_{SR}\}$ of $\{A_S\}$ such that the corresponding subsets $\{\vec{a}_{SR}{}^\dagger\}$, of $\{\vec{a}_S{}^\dagger\}$ used in equations of the type (2), will suffice for a nonambiguous estimation of $\vec{a}$ and $\vec{p}$. (This process may yield new estimates also for those elements of $\vec{a}$ for which the estimates $\vec{a}_{SR}*$ were available.) The estimates a_S* which one obtains in this process (for any elements of $\{A\}$) are considered to be "on the system $\{a_S{}^\dagger\}$". This definition attempts to give a precise description of current usage. It is clear that the same measurements $\vec{x}*$ will lead to different sets of estimates $\{\vec{a}_S*\}$ when the model function Ψ (or Φ), delete or the way in which the subset $\{\vec{a}_{SR}{}^\dagger\}$ is selected from $\{\vec{a}_S{}^\dagger\}$, or both, are open to choice.

3. RESTRICTING THE AMBIGUITIES

One might think that there should be little question about ambiguity in the model function, since relationships between the variables are determined by the physical and geometrical circumstances of the situation, including the number and kind of relevant parameters. Remember, however, that any model function is in reality always only a more or less accurate representation of the actual physics and geometry of any situation; witness the fact that linear (or quadratic) interpolation is based on the assumption that in certain restricted circumstances, a linear (or quadratic) relationship between the relevant parameters is for the purpose at hand a sufficiently accurate representation of their relationship. Thus, different - equally defensible - choices for the model function will then lead, - even if nothing else is changed - to different sets of $\vec{a}_S$*. The only way to relieve this undesirable ambiguity is for all investigators of any particular situation to agree on the same model function. This is obviously not likely to happen. Even more so, there will be different sets of parameters $\vec{p}$ associated with different models. The relationships between the sets of parameters belonging to different models is difficult.

Also, - even for identical models - the set(s) of estimates $\{\vec{a}_S^*\}$ (and $\{\vec{p}^*\}$) which are obtained as the result of the investigation will apparently differ for different sets $\{\vec{a}_{SR}^\dagger\}$. The $\{\vec{a}_S^*\}$ and the $\vec{p}^*$ will generally be functions of other, fixed parameters which are not adjusted in the course of the calculations. A set of ranges of these parameters defines a region, such that only such components of $\vec{a}_{SR}^\dagger$ whose fixed parameters are within the region in consideration are chosen for a particular adjustment. Changes of the region that become necessary will thus lead to discontinuities in the $\vec{a}_{SR}^*$ and the $\vec{p}^*$, unless a weighting function is applied in such a way that at the edge of a region, an element of the reference set enters (or leaves) the adjustment with weight zero at the boundary of the region and that, once a region is defined, all elements of $\{\vec{a}_S^\dagger\}$ whose parameters fall within it are included in the adjustment.

Such weighting could, for example, be accomplished as follows. Let q be one of the parameters which defines the region that determines the set $\{A_{SR}\}$, such that all elements of $\{A_S\}$ for which q satisfies $q_0 \leq q \leq q_1$ are included in $\{A_{SR}\}$. The weighting function with which an element for which the parameter in question has the value q enters the adjustment could, for example, be

$$w(q)=\left(\frac{4(q-q_0)(q-q_1)}{(q_0-q_1)^2}\right)^2 \tag{4}$$

which vanishes together with its derivatives at the boundaries ($q=q_0$ and $q=q_1$), and reaches a maximum of 1 at midinterval ($q=\frac{1}{2}(q_0+q_1)$). If the weight of an element of $\{\vec{a}_{SR}^\dagger\}$ in the adjustment within the region which determines $\{A_{RS}\}$ were specified by Eq. (4) (the generalization to more than one variable q is obvious), the estimates $\vec{a}_S^*$ for the target variable and $\vec{p}^*$ for the reduction parameters would vary continuously and smoothly when this region is changed by a continuous change of its boundaries. If investigators in various areas could agree on standardizing modeling functions and sampling regions, and would adopt one of the several feasible weight functions, the presently existing ambiguities in establishing data estimates on a particular system would be removed. *Until this happens, there can be no precise meaning of the unqualified statement that certain data are on a certain system.*

4. REDUCTION UNCERTAINTIES

Any system consists of a set of estimates. The process by which any estimates are obtained yields, at least in principle, also their covariance matrix Q. (Unfortunately, Q is usually not communicated.) The variance of any function $\phi(\vec{a}^*)$ of the set $\{\vec{a}^*\}$ of estimates is given by $\sigma_\phi{}^2=(\partial\phi/\partial\vec{a}^*)^T Q(\partial\phi/\partial\vec{a}^*)$, that of a function $\psi(\vec{a}^*, \vec{y}^*)$ of these estimates $\vec{a}^*$ and another, with them *uncorrelated* set $\{\vec{y}^*\}$ of statistical variates (e.g., measurements) whose covariance matrix is Y is given by $\sigma_\psi{}^2=\sigma_a{}^2+\sigma_y{}^2$ with $\sigma_a{}^2=(\partial\psi/\partial\vec{a}^*)^T Q(\partial\psi/\partial\vec{a}^*)$ and $\sigma_y{}^2=(\partial\psi/\partial\vec{y}^*)^T Y(\partial\psi/\partial\vec{y}^*)$. The standard error σ_ψ is a certain fraction of the most likely amount by which the function ψ, evaluated at its arguments, will deviate from the "truth"; it is, in fact, inversely proportional to the "average" accuracy of the function ψ. (This fraction depends on the distribution function of the errors involved, for a normal distribution it is about 3/2).

Common usage regards relative estimates, $\vec{a}^*$ obtained from solving a set of equations of the type (2) with respect to $\vec{a}$ (and $\vec{p}$), to be "*on the system*" of $\{\vec{a}_R{}^\dagger\}$. This somehow implies that there should be no systematic differences between two different sets $\{\vec{a}_1{}^*\}$ and $\{\vec{a}_2{}^*\}$ of estimates of the same set of quantities as long as $\{\vec{a}_1{}^*\}$ and $\{\vec{a}_2{}^*\}$ were obtained with the use of the same $\{\vec{a}_R{}^\dagger\}$. Such differences are, however, in fact unavoidable. In order to demonstrate this, we regard two different sets $\{\vec{x}_1{}^*\}$ and $\{\vec{x}_2{}^*\}$ of measurements[1], not necessarily of the same quantities. These are used for the estimation of two sets of estimates $\{\vec{a}_1{}^*\}$ and $\{\vec{a}_2{}^*\}$ of $\{\vec{a}\}$ from adjusting for $\{\vec{a}_1{}^*\}$ and $\{\vec{p}_1{}^*\}$, and $\{\vec{a}_2{}^*\}$ and $\{\vec{p}_2{}^*\}$, respectively, the equations

$$\Phi(\vec{a}, \vec{a}_R{}^\dagger, \vec{x}_1{}^*, \vec{p}_1)=\vec{0}$$

and

$$\Psi(\vec{a}, \vec{a}_R{}^\dagger, \vec{x}_2{}^*, \vec{p}_2)=\vec{0}$$

respectively. From the adjustment based on $\Phi=\vec{0}$ one will get Q_1 and from that based on $\Psi=\vec{0}$, we get Q_2 as the covariance matrices of $\{\vec{a}_1{}^*, \vec{p}_1{}^*\}$ and $\{\vec{a}_2{}^*, \vec{p}_2{}^*\}$, respectively. The covariance matrices P_1 and P_2, respectively, of $\{\vec{a}_1{}^*\}$ and $\{\vec{a}_2{}^*\}$, respectively, are, of course, minors of the corresponding Q_v. The systematic difference $|\phi(\vec{a}_1{}^*, \vec{y}^*) - \phi(\vec{a}_2{}^*, \vec{y}^*)|$ can always be directly computed. Its most likely value is the abovementioned fraction of the dispersion of this difference (as far as it depends on the differences between the different estimates for the same $\vec{a}$) which is

$$[(\partial\phi/\partial\vec{a})^T(P_1+P_2)(\partial\phi/\partial\vec{a})]^{1/2},$$

clearly not zero. This shows that in fact, somehow, the two different estimates $\vec{a}_1*$ and $\vec{a}_2*$, in spite of entrenched terminology, do not represent the same system. Since one of them ($\vec{a}_1*$, say) could be $\vec{a}_R^\dagger$ (in this case, $\vec{a}_2*$ would be simply a reestimation - possibly for the intended purpose of improving precision and accuracy - of the values $\vec{a}$ for the reference set) we see immediately that the "*preservation of the system*", is in the sense a system was defined in this paper as *impossible* a task as is the construction of a perpetuum mobile. This fact should cause investigators to reexamine the often followed practice of carefully "reducing different sets of observations first to the same system" and leaving the determination of the errors (i.e., estimation of the differences against the true values) to a common, later step. The assumption that the system values and the reduced values have (by and large) the same "systematic errors" may well turn out to be an illusion.

5. CONCERNING ACCURACY

Relative estimates of a certain set of target variables can actually be more accurate than those which constitute the reference system to which they were adjusted. In order to show this, we suppose the elements of a system $\{\vec{a}_R^\dagger\}$ were estimated absolutely by solving for $\vec{a}$ (and $\vec{p}$) the equations

$$\Phi(\vec{a},\ \vec{x}*,\ \vec{p})=\vec{0}. \tag{5}$$

Let Q_a be the covariance matrix of $\{\vec{a}_R^\dagger\}$ which resulted from this adjustment. Now obtain a set of relative estimates $\{\vec{a}*\}$ of the same variables from solving for $\vec{a}$ (and $\vec{q}$) the equations

$$\Psi(\vec{a},\ \vec{a}_R^\dagger,\ \vec{y}_1*\ \vec{q})=\vec{0} \tag{5}$$

where $\vec{y}*$ are certain measurements and $\vec{q}$ certain parameters. Consider a function $\phi(\vec{a})$. Denote the variances of $\phi(\vec{a}_R^\dagger)$ and $\phi(\vec{a}*)$ by $\sigma_\dagger^2$ and σ_*^2, respectively. It is now quite feasible that, given such a pair $\{\vec{a}_R^\dagger\}$ and $\{\vec{a}*\}$, $\sigma_\dagger^2>\sigma_*^2$ for most ϕ, indicating that the estimates $\{\vec{a}*\}$ are more accurate (there is no question that they can easily be more precise) that the $\{\vec{a}_R^\dagger\}$. This seems strange at first, because one is tempted to argue that the errors introduced by the measurements $\vec{y}*$ are added, in a way, to the errors originating from the $\vec{a}_R^\dagger$ so that the errors of the $a*$ must be at least as large as those of the $\vec{a}_R^\dagger$. Consider, however, that Eq. (6) may, - while incapable of determining $\vec{a}$ without $\vec{a}_R^\dagger$ - subject the $\vec{a}$ to constraints not enforcable by Eq. (5) because of the way in which the $\vec{x}*$ were obtained. Consider, for example, a set $\vec{a}_R^\dagger$ which are available just as a series of pairs of numbers, while new measurements $\vec{y}*$ allow one to enforce the constraint that the number pairs which constitute the $\vec{a}$, if interpreted as coordinates in a plane, must lie on a straight line. This powerful new constraint may well make the relative estimates $\{\vec{a}^*\}$ "on the system of the $\{\vec{a}_R{}^\dagger\}$" more accurate than the elements of the system. Depending on the circumstances, there may or may not be significant systematic differences between $(\vec{a}_R{}^\dagger)$ and $(\vec{a}^*)$; examples for both these possibilities could be quoted from the literature. In any case, however, these considerations indicate that the imposition of newly available constraints through additional sets of pertinent measurements may well generate a new system consisting of relative estimates which is more accurate and precise than the original system of absolute estimates to which the new system was reduced.

Which of two concrete systems is more accurate can be decided only by calculating the covariance matrices of the estimates which constitute the systems. This is usually connected with a large computational effort and seldom, if ever, properly done. Therefore, one finds frequently misjudgments on the basis of incomplete information.

NOTES

1. One might also define a measurement as an estimate that is directly obtained by reading the scale of a measuring device without any calculation involving the estimation of additional parameters. Note that according to this definition, individual readings on different settings of an experimental setup on the same object, which will differ from each other only because of the measuring errors, are measurements, while their average is not, because of the measuring errors, are measurements, while their average is not, because it is an estimate computed by taking the mean of different estimates of the same quantity.

*Contributions of the Department of Astronomy of the University of Florida #38.

DIGITAL IMAGE PROCESSING AND DATA RETRIEVAL : THE NEEDS

A. Bijaoui
Nice Observatory - France

During the last decenny, Digital Image Processing (D.I.P.) has been introduced in astronomical studies to allow the information extraction.

In a first step, D.I.P. has been used essentially to provide enhanced images (noise reduction, deconvolution, contrast enhancement), to reduce geometrical or photometrical distorsions and to extract rough data. So, a few reference date are needed (some comparison lines for example).

As the D.I.P. allows us to fit all kinds of images with a model, an important software has been developed to obtain values of positions, magnitude, diameter, equivalent width, radial velocity... But to transform rough data provided by the adjustments into intrinsic astrophysical quantities, we need to introduce some data from catalogues, tables...

In the case of spectrography, there is generally no difficulty to identify comparison lines, but it arrives in our center that some difficulties arise from the nature of the spectral comparison source, the wavelength field... At present it is the user who identifies these lines, but a catalogue of good comparison lines according to the set of conditions (spectrograph, lamp, exposure) is a first need for D.I.P.

If we use some well-known tables to identify stellar lines (Solar spectrum, Moore tables ...) a second need for D.I.P. lies in a catalogue of lines according to the spectral type, allowing us to extract, for a defined dispersion and wavelength field, the lines corresponding to the star studied, leading to interactive or automatical identification. A similar need exists for spectra of nebulae or galaxies.

To allow us to go further in data reduction, catalogues of data provided by models, can be built.

For direct photography, the use of the C.S.I. and the data provided by the Strasbourg Stellar Data Center allows us to do direct astrometric

C. Jaschek and W. Heintz (eds.), Automated Data Retrieval in Astronomy, 109–110.
Copyright © 1982 by D. Reidel Publishing Company.

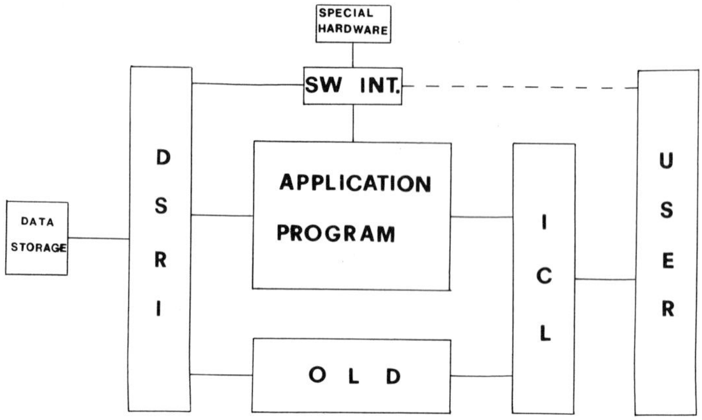

Fig. 1 - Interactive application oriented software environment

a) Flexible access to all facilities of the DP system. The interface between the user and the application module can be implemented advantageously through an Interactive Control Language (ICL)

b) Transparent access to local and remote data storage subsystems. The interface to local data storage requires an effective standard organization of the data structures and/or the definition of the organization of any remote data storage sub-system or support accessed. An automatic log of any operation through the data storage interface is necessary to trace and maintain the documentation of the DP task performed.

c) Transparent access to special purpose components. The interface to specific sub-systems (graphics, array processors, etc...) must allow component-independent access to available facilities through a proper design of the interface.

d) Effective integration of new application modules. The integration interface must provide on-line self documenting standards for entering any new modules.

Notice that all interfaces include necessarily a section which is operating-system dependent. Since this section can only be minimized it follows that full standardization of application software is possible exclusively within a common hardware environment. This limitation will not apply to application modules if a careful set of definitions is agreed on the sections of the interfaces to the application module facing the module itself.

3. A TENTATIVE INTEGRATED SOLUTION

We used the considerations reported above to implement an integrated solution dedicated to astronomical applications. This solution was implemented and run on a DEC PDP 11/45 computer under a RSX 11 M operating system. It operates according to the rules given above through

- straightforward interface to the user
- simple interfacing rules to the application modules
- simple maintenance and updating rules with on-line documentation for the ICL
- self-documentation and eventual batch operation of any successful session
- standard organization of data storage through standard structures of data files and standard user-transparent R/W routines. High efficiency was achieved by means of fixed and I/O optimized lenght random access file records, equivalence of physical and logical records, self-consistent file header, auxiliary user and system accesible storage space. The data organization includes an archive facility.
- self contained user-transparent access to a colour graphic subsystem AYDIN 5214A of 512x512x3 or 256x256x6 format.

The preliminary results show that also a very simple solution (if correctly designed and integrated into the available hardware), is able to give

1. basic homogeneity and compatibility of user programs, easy maintenance and access to data archives and full exchangeability of data between different users due to standardization of the data storage organization ;

2. easy access to all DP facilties for end-users and straightforward access to complex graphics with minimum overload and overhead to the application program.

3. CONCLUSION

We know that the tentative solution quoted here can be improved largely and that it is certainly very far from an optimum solution to the problem of standardization of astronomical software. We believe however that the experience acquired during this work may be useful to the astronomical community because it was made working in an astronomical environment on astronomical data taking into account the actual astronomical DP scenary.

We make ICL including AYLIB colour graphics available to the astronomical community and invite everyone interested in the standardization problem to push for and participate in the definition of international standards for astronomical applications.

the higher density with the longest possible physical records (blocks), thereby greatly reducing the number of tapes while simultaneously decreasing computer processing time. All tapes received are therefore processed by a special program to reblock them to 6250 bpi with variable block size and spanned logical records (VBS format).

Table 1 demonstrates the advantages of processing the data for storage and retrieval purposes. The VBS record format results in a decrease of about a factor of 10 in the input/output computer time when data are retrieved to prepare requester tapes. The significantly fewer

Table 1. Comparison of Original and Reblocked Tapes

	Original Tapes	Archive Tapes
Density	1600 bpi	6250 bpi
Blocking	Unblocked	Blocked 32,000 bytes/record
Record Format	Variable	Variable, Spanned
Number of Tapes	3084	310

number of archive tapes also greatly simplifies the retrieval of images since fewer tapes need be read for recovery of a given number of files. More importantly, due to limited storage space for tapes at our computer facility, archive tapes must be stored at another location and sent to the computer area to be processed; hence, the physical handling of tapes and requirements for storage at the computer center are markedly decreased.

II. THE DATA MANAGEMENT SYSTEM

Even with the fewer number of tapes resulting from reblocking of the data, it is obviously of great importance to use a computerized list of images to locate data to be retrieved from the archive tapes. Maintenance of an image indexing data base has proved to be a significant part of the IUE task, since we require the index to contain enough information for cross checking to locate mis-specified and incomplete identifications on requests (several errors have been discovered in this way). Until recently, the indexing data base was updated manually, with the usual error and identification problems, especially for images which had been reprocessed and supplied as replacements. Also, the archiving and project facilities maintained independent data bases containing considerable amounts of redundant information and software for data base management. With the development by the IUE project and observatory of more sophisticated management and housekeeping routines, it was decided that the data bases should be merged, thus allowing the use of previously developed software for editing and automatic entry of new records to the unified data base.

Upon completion of each high-density archive tape, a program is run which determines the locations of all images on the tape. This information is written to a disk data set in the form of commands to be used by a subsequent background editing program which updates the data base. The disk data set is quality checked at an interactive computer terminal and added to another disk data set, which is actually used as the input to the editing program. After the editing run has been made, the input data set is reinitialized and ready for additional entries for newly completed archive tapes.

A special accounting program, previously used to monitor the IUE project data base, has been expanded to analyze and compute statistics for NSSDC information in the merged data base. Weekly reports are now produced which list images delivered to NSSDC but overdue for processing or shipment to the European data centers.

III. SUMMARY

The archiving and indexing of IUE spectroscopic data at NASA has progressed from a largely manual operation to almost an entirely automatic procedure through the development of processing and data management software. At present the automated system is used only for data taken at the NASA IUE Observatory, but efforts are now underway to create an automated data base system for VILSPA data. The new system has proven very effective in eliminating most errors and increasing the efficiency of our retrieval and dissemination efforts. The improved indexing and decreased turn-around time for data availability will provide faster and easier IUE data accessibility for research and analysis purposes.

An additional possible improvement for data processing of observations taken in a guest-observer environment such as IUE, where processed data are transmitted to each observer for further processing and analysis at one's home institution, would be to adopt a standard format for data transmission. The current IUE image-type format, although well suited to the information content, consists of mixed mode IBM binary and EBDIC data which are sometimes difficult to read and process by other computer systems. The use of a flexible format such as FITS (Wells and Greisen 1979), which has been designed specifically for the transport of image data to other installations, would greatly facilitate data handling by the astronomers themselves, an obviously very important aspect of a guest-observer structured program.

REFERENCE

Wells, D.C., and Greisen, E.W. 1979, in *Image Processing in Astronomy*, Ed. G. Sedmak, M. Capaccioli, R.J. Allen, Osservatorio Astronomico di Trieste, p. 445.

leads to new developments in information management and organization methods. Four main lines of impact are due to:

- new orientations in the structuring of information
- developments in the area of data management in a distributed environment
- innovations in teleprocessing
- attempts to integrate structured, textual, and graphic information

These developments raise planning and organizational issues which have to be settle in order to achieve coherent development of the information processing facilities within the organization.

3. Technical methods and tools for systems development. New hardware and software architectures can help us obtain our objectives of controlling the costs and improving the quality of computer-based systems. New algorithms and data structures can sometimes have tremendous impact on the quality of those systems. There is also a need for new technical methods and tools that will permit us to use effectively whatever architectures are available; in particular methods and tools for analysis, design, construction, and testing of computer-based systems. The main reasons, why we must look up for new techniques are:

- insufficient system quality (e.g. reliability, maintenability, portability)
- difficulties in finding out, what the users and purchasers of a system really need
- low productivity of development personnel

4. Economic, legal, and social implications. A greater awareness of economic, legal, and social aspects of data processing is needed. Economic impact of an introduction of new systems on an organization have to be carefully investigated. There is a need for the protection of applications and data (present procedures seem to be insufficient to handle sensitive and confidential data). Social implications of data processing have to be reconsidered (e.g. rationalization effects on on employment).

There is a necessity to integrate above mentioned means into the "optimal solution" of effectiveness problem. The major integration difficulties arise from:

- scope and complexity of these aspects
- underestimation, misunderstanding, and underweighing of some aspects
- heterogeneity of disciplines involved

SOFTWARE MODELS FOR REAL-WORLD APPLICABILITY TESTS

As one possible solution to the problem of effectiveness of data processing, the development of flexible software models is proposed. The rational for such an approach is the possibility to test these models for applicability in the real-world environment, e.g. not only technical characteristics of the particular model are being tested but also functionality thereof. Rather straight-forward method is being proposed:

- indentify needs
- specify requirements
- develop a model as a laboratory system
- test for technical and functional qualities

Iterate above steps as many times as it is feasible.Then develop a product.

The crucial step in the method is to develop a model (with limited design resources); this model has to serve as a laboratory system for the development of a specific and operational system. Hence the minimal characteristics of such a model should be:

- portability
- utilization of commercially available software-components
- serving as a laboratory system for
  - further development of system functions
  - testing of system characteristics (e.g. response-time, user-friendliness)

REFERENCES

Mistrik, I. and Koch, W., "Project Plan: Development of Software Model for Decentralized Information Procurement and Processing on Minicomputers and Microprocessors". Gesellschaft für Information und Dokumentation mbH, January 1981.

Mistrik, I. and Koch, W., "Dedicated Information Processing System: Development of a Software Model". In Proceedings of Annual Conference of EUSIDIC, Bern, Switzerland, October 1981.

the CSI and the SAOC. As a result of this comparison, numerous problems with SKYMAP were detected. Thus the simple process of extracting data became quite tedious since, even though the preliminary comparisons were done by computer, many checks had to be done by hand. A preliminary subset of stars was extracted from SKYMAP by applying the four criteria. Many stars lacked spectral class information. So the CSI was examined to see if missing spectral classes could be substituted. During the comparison we detected numerous SKYMAP stars which did not appear in the CSI. Since the CSI is believed to be complete down to $m_v = 9.5$ (Ochsenbein et al., 1977), this was very disconcerting. Our analysis revealed that many of the problems were due to duplicate SKYMAP entries; i.e., a star appearing in both the CD and CPD might have each entry in SKYMAP. Another problem was uncovered when SKYMAP was compared with the SAOC. It was discovered that numerous SKYMAP positions differed from the SAOC positions by more than half a second of arc. This led to a thorough revision of our subset. All SKYMAP stars whose positions did not originate in the SAOC or the AGK3 were eliminated. SAOC and AGK3 positions, proper motions and tabulated uncertainties were substituted for SKYMAP data at epoch 1950.0. In the end, every star in the NSWC subset needed some correction. After all corrections were completed, the catalogue criteria were applied again and the final NSWC Master Catalogue (NMC) of 43,075 stars was produced. The NMC has a star density of approximately one star per square degree, although the distribution is not uniform. We feel confident we have compiled as complete and accurate a catalogue as possible. Any errors detected in the source catalogues have been reported to the authors and to the distributing data centers.

Vehicles such as LANDSAT-D will experience rotation rates defined by an active attitude control system about an axis normal to the orbital plane. If nominal position, velocity and orientation are known, the approximate motion of the star sensor's field of view (FOV) across the celestial sphere can be predicted. In order to determine precise attitude at particular times in the mission, we must be able to rapidly retrieve stars from the catalogue. Yet we do not want to be forced to search the whole NMC every time. Thus based upon specific mission parameters, we can extract a subset of NMC stars and associated data. Such a subset is known as the mission catalogue (MC). To form the MC, we used the following criteria.

a) Eliminate stars violating mission criteria.
b) Only include information necessary for star identification.
c) Store data in a way which is quickly and easily accessible.

The first two steps in forming the MC involve determining which stars to eliminate and what information to retain. This decision is based on such mission-specific factors as instrumental characteristics, vehicle constraints and software restrictions. Once the MC subset is selected, the stars are placed in cells which are treated as elements of a random access file. Software constraints determine the dimensions and number of cells. Globally, there are 1820 cells in our simulations. Those within $\pm 50^{o}$ declination (δ) zone have dimensions of $5^{o} \times 5^{o}$. For higher δ zones, the right ascension (α) dimension increases until a cap terminates each hemisphere. An indexing scheme allows each cell to be uniquely identified. An algorithm converts the α and δ ranges of each cell to a single integer identifier. Consequently, any star's α and δ can be converted to a cell identifier and its information can be stored in that cell of the random access file. Once a MC has been created for a particular mission, the MC can be used for all observations during the mission unless mission criteria change.

The star identifier software needs a method of quickly accessing a portion of the MC. The vehicle rotation causes the FOV to coarsely trace a swath parallel to the orbital plane across the celestial sphere. The swath length is dependent on the duration of an observation. At specific time increments during observation, the coordinates of the FOV center can be determined and consequently the MC cell containing this point can be identified. Due to uncertainties in the true celestial coordinates of the FOV center, the stars in the FOV cell and in the surrounding cells must be extracted from the MC. These stars can be quickly retrieved from the random access file once the cell identifier is known.

In the next step, the stars are ordered in a sequence more easily manipulated by the software and further eliminations are performed. This is accomplished by transforming the celestial coordinates to a frame having the polar axis perpendicular to the orbital plane. The swath limits including tolerances for pointing uncertainties are easily defined in this frame and any stars exterior to the swath boundaries are eliminated. A large fraction of the extracted stars can be eliminated in this process. Before the remaining stars are sent to the star identifier, their positions can be updated to the instant of observations. If necessary, we only need to apply precession, nutation, aberration, parallax and/or proper motion to a small subset of the MC stars. This access scheme has proven to be quite efficient in simulated missions. Although this paper addresses a specific concept, these techniques could be generalized to solve other aspects of astronomical data retrieval.

REFERENCES

Gottlieb, D.M.: December, 1977, Computer Sciences Corporation, CSC/SD-76/604/UD1, Skymap System Description (Revision 1).

Ochsenbein, F., Egret, D., and Bischoff, M.: 1977, "The Catalogue of Stellar Identifications" in Compilation, Critical Evaluation and Distribution of Stellar Data, pp. 31-36, C. Jaschek and G. A. Wilkins, Editors, D. Reidel Publ. Co., Dordrecht, Holland.

order of 1% and S-distortion will be around 5%. So, at the edges of the field of view, the pictures will be distorted by a few tens of picture elements (pixels).

Each pixel contains a number, called a count, which is related to the number of photons incident on the corresponding part of the detector faceplate. This count, however, may be non-linear, especially at high incident photon rates (i.e. at high count rates).

Ground and in-orbit calibration of the instrument will permit the count values to be corrected to within about 1%. Pixel to pixel accuracy will be achieved using ground based results, but accurate absolute calibration will have to await in-orbit observations of reference celestial objects.

The routine, geometric and photometric correction of FOC images will be carried out with software produced as part of the FOC development. This software package will correct for the known imperfections of the FOC and will produce images calibrated to a linear amplitude on a rectangular grid.

In addition to the standard processing described above the astronomer may want to carry out a number of other corrections to the data which however depend on the precise nature of the scientific information being sought. These include for example pixel overflow correction, particle events correction, removal of internal reflection "ghosts" and removal of known blemishes and fiducial marks. Special software will also be required for the various non-imaging modes of the FOC. In the spectroscopic mode there will be a requirement for transformation to wavelength scale, correction of the curvature of the spectrum and precise determination of the position of the slit in the sky. In the objective prism mode a wavelength scale transformation and location of the sources in the sky will be needed.

After completing the tasks described above the astronomer is only at the beginning of the process of scientifically analysing his data, and he may wish to perform a number of additional operations on his data, for example, he may wish to fit model curves to surface brightness profiles in order to determine standard parameters such as "stellar" nucleus size, disk scale length, etc.

To support these routine and non-routine corrections and the task of scientific data analysis operations a basic set of general purpose software will be required. This will include modules that will allow operations to be carried out such as, scaling; panning; blinking; rotation; translation; zooming; adding, subtracting, multiplying, and dividing images by other images; overlaying images; displaying images with different colour schemes etc.

THE ESA FACILITY

In order to allow development of the software described above ESA is establishing a dedicated facility for the FOC in ESTEC. The hardware is based on a VAX 11/780 computer with 1.5 M bytes of memory, two 67 M bytes disk drives, two magnetic tape units (800/1600 bpi), a line printer, two VT100 terminals, a floating point accelerator facility and the FORTRAN IV plus compiler. In addition a colour image display system, the Sigma ARGS, a quick-look hard copy facility and a TV-type camera facility for digitizing photographs are being purchased. A contract has been placed for the development of the system software and major utility programmes which will provide the "skeleton" for the scientific applications programmes.

The FOC Image Processing System (FIPS) software environment provides a user interface designed for the astronomer and a set of system software interfaces for the development and integration of application software. The components that are provided to support both the applications programmer and the end-user are as follows:-

- a systems library of primitive routines for access to image data, primitive routines for access to user supplied parameters and for access to the ARGS.
- a user support package which includes centralized dialogue, command interpretation, also including high level language interpretation.
- an image processing package to provide the basic image processing functions.

FIPS runs under the supervision of the VAX/VMS operating system. The structure is that of multi-processes, whereby each (sub)process communicates with a central control process called gateway which conducts the dialogue with the user. Each user will get his own copy of gateway. User input is decoded and placed in a fixed data array where it can be easily accessed by the gateway process. Gateway is responsible for calling the desired processes into core and transmitting the parameters to the created process. Furthermore gateway accepts, validates and maintains user identification which is used for identification of and protection against corruption of images. Via the gateway process it is possible for the user to move his attention from one programme to another while these run simultaneously (multi-processing). However, the sequence in which programmes can be served with user input is determined by the programmes rather than the user. Gateway also produces a comprehensive log of all dialogue for the user, which is automatically printed at the termination of the session.

The FIPS development is expected to be completed towards the middle of 1982 at which time it will be available to any interested astronomer.

REFERENCES

1. Macchetto, F., Pacini, F., Tarenghi, M., (eds.), Proceedings of the ESA/ESO Workshop on Astronomical Uses of the Space Telescope, Geneva, 12-14 February 1979.
2. Macchetto, F., van de Hulst, H.C., di Serego Alighieri, S., and Perryman, M.A.C., The Faint Object Camera for the Space Telescope, ESA, SP-1028, October 1980.

II. DATA SET MODIFICATIONS, ADDITIONS AND PROCESSING

The basic procedures used for verifying and checking data in newly acquired and produced catalogs have been described by Underhill *et al.* (1977) and Hill (1981). Although certain errors can be discovered by the computer checking of data for consistency, many potential problems for data processing and application can only be uncovered by actually working with and having a basic familiarity with the data.

Many older catalogs, which were computerized prior to the development of sophisticated data processing techniques, were produced with very restricted formats, or certain data fields were omitted to confine the data records to 80-column computer card images. For such catalogs, we usually add missing data fields to complete the records as published, and add remarks or reference files where necessary. Data received in print rather than processing format are converted to structures more appropriate for analysis and plotting.

III. ASTRONOMICAL DATA CENTER OPERATIONS

It is found extremely important, although time consuming, to create and maintain detailed records on all facilities required and every product distributed with each requested catalog. In addition to providing information necessary for cost evaluation and funding purposes, these detailed records are often referred to later as new editions of particular catalogs are received or additional related information becomes available, since all recipients of a catalog should be notified as new products are introduced. The distribution of catalogs from the NSSDC is a great advantage in this regard, since existing data management systems provide monitoring for all request activity. Upon receipt of a request for a specific catalog, a record is entered into the system, a projected completion date is registered, and the request is monitored via monthly reports until completed. At the completion of a request, all computer time used and each product distributed are coded on completion forms and entered into the system. This system also produces annual reports of all request activity and products disseminated. A separate file contains records of all forms of data products associated with each catalog and all related materials to be distributed with a tape or microform version of a data set, such as information pages, error lists, related papers from the literature, and special documentation.

IV. DOCUMENTATION, STATUS REPORT AND SPECIAL SERVICES

Since the initiation of our work with astronomical catalogs, detailed documentation has been a very important aspect of our activities, and it continues to demand a substantial amount of time. A standard format has been used for all documents with only minor modifications depending upon the catalog being described. Each document contains five sections designed to present a general description of a catalog, the record format, the tape characteristics, remarks, modifications and references, and a sample listing of data exactly as they appear on the tape. We consider the section on remarks and modifications to be extremely important and attempt therein to give a complete history of all changes made with respect to the original published catalog, plus the source of the initial machine-readable version.

It is important to provide information concerning the status of each catalog on file. We maintain a disk file containing current status information on every catalog, so that users know not only which data sets are in revision and/or are not available at any given time, but also the data quality and reliability of any catalog that they request. The above data set is also used to produce status reports which are published in the *Astronomical Data Center Bulletin*.

If resources permit, a data center should provide special services to the user community. These services involve mainly the development of software to process various catalogs to generate specialized output. We have developed a variety of services such as computerized searches of data and bibliographic catalogs (Mead *et al.*, this volume), preparation of subsets of and specially sorted catalogs, and generation of plotted output for finder charts and overlays (Nagy *et al.* 1980).

V. SUMMARY

A data center should be active in attempting to improve the quality and homogeneity of data distributed. For the archiving and distribution of astronomical catalogs, this requires the employment of several astronomers expert in the area of data processing and at least basically familiar with most of the data disseminated. Detailed documentation containing a byte-by-byte format description, correction history, and sample listing should be prepared where necessary. Special services for data searches and processing should be provided for users not having those capabilities. The standardization of data structures and formats for groups of related catalogs would greatly facilitate the development of data services and special processing software.

REFERENCES

Hill, R.S. 1981, *Astron. Data Center Bull.* 1, 69.

Mead, J.M., Nagy, T.A., Hill, R.S. and Warren, W.H. Jr., this volume, 9.

Nagy, T.A., Mead, J.M. and Warren, W.H. Jr. 1980, *Astron. Data Center Bull.* 1, 3.

Underhill, A.B., Mead, J.M. and Nagy, T.A. 1977, in *IAU Colloquium No. 35, Compilation, Critical Evaluation and Distribution of Stellar Data*, ed. C. Jaschek and G.A. Wilkins (Dordrecht: Reidel), p. 105.

Wilkins, G.A. 1977, in *IAU Colloquium No. 35, Compilation, Critical Evaluation and Distribution of Stellar Data*, ed. C. Jaschek and G.A. Wilkins (Dordrecht: Reidel), p. 287.

common programming languages (CPL), such as FORTRAN and ALGOL 60 or even PL/I, ALGOL 68 and PASCAL, have had their golden age about ten years ago. During the last decade programming practice has brought in use new programming tools - namely data bases (DB) and application program packages (APP) - both with appropriate control systems and input languages. These languages are often not compatible with CPL which are still more appropriate for writing new algorithms. On the other hand DB and APP are more suitable for storing and using previously accumulated information and knowledge.

In the system SPORA all these means - DB, APP and CPL - are tied together and form a (hopefully) balanced integrity.

One more concept of the 70-ies is that of abstract data types (ADT). In contrast with DB and APP this concept has come not from practice, but rather from the intention of language designers to raise the level of programming languages, to make these languages closer to natural languages and simpler for use and understanding. Therefore this concept was accepted and widely used in the system SPORA and in its input language "Descartes".

There are two kinds of abstract objects in the system : an abstract type and an abstract map (or mapping). An abstract type may be either a primary one or a relation.

A primary type is introduced by a declaration, an example of which is given on fig. 1. The declaration contains the name of the type, the names and types of operations that can be performed on values of the type, and the axiomes expressing the properties of these operations. In the example identifiers calend and julian are names of so called representations. Each representation is bound with a method of physical measurment of a value of the type. Each physical representation may in its turn allow different implementations in various programming languages. The properties of a representation can be described, e.g., as in fig. 2.

We cannot go into details of the programming implementation description which tell the system how to denote a value of the type, to duplicate it and to convert it from one physical representation into another. This part of the type declaration is denoted by "..." in the figure.

An example of a map declaration is shown on fig. 3. Again the implementation details of this declaration are omitted.

We distinguish between relations of a data base, which are called tables, and relations of an application program package, which may have more complex (hierarchical) structure. Let us consider closely only the first kind of relations.

The typical example of a table declaration is given in fig. 4. It contains a table name, names and types of columns, a unique key of the table, and possibly some indications of more efficient ways of access to the table.

The next concept is that of a data base schema (fig. 5). A data base schema describes tables that are stored permanently in this data base. For each specific program it is possible or even necessary to describe a subschema of the schema (fig. 6). The names of tables and the names and types of their columns must be the same as in the data base schema. The number and order of the columns may however be different.

One possibly feels that the language of such declarations is too complicated, especially in the part dealing with representations. It may be, but these declarations have to be written only once for a relatively long period. This has to be done by an advanced user. We call him (or her) a designer in contrast with the broader range of users. Their task is to describe a specific problem by means of one or more Descartes statements.

The main part of a statement is a request, which consists of a projection, a line description and a condition. Only the line description is obligatory. In a request, presented on fig. 7, we ask for an intermediate table, containing observations of Mercury at third contacts made during an eclipse in 1973 having the maximum value of O-C.

Each statement is to be placed in a program, written in some host (base) programming language. In the case of the examples given in fig. 8 and 9 this language is ALGOL 60.

The part of a program presented in fig. 8, calculates the average value of O-C for all observations contained in the personal archive of a user. The fragment, exemplified on fig. 9, recalculates those values of O-C, which do not lie within interval [C1, D1] . The construction r.oc → cvt julian (A) bounds for each line r of the table personal-archive the value of the attribute oc in the representation julian with the host language variable A. During the execution of the statement B : = B + A (fig. 8) this variable assumes that value.

A user must have only a basic knowledge of host programming language - how to declare variables and to describe the simplest calculation.

We have not mentioned here another part of the language "Descartes" that deals with application program packages. Using this part a problem can be stated even more simply and on a more abstract level.

The general structure of the system is shown on fig. 10. The model of an application area is created by designer. The user specifies his problem.

A problem is always stated in one specific area model. There may be many models in the system and many problems per model. A model defines terms used in its area and their connections and properties. All the

```
db observations ;
    dbtab mercury transit
      (nobs : observation number ; nc : contact number ;
       dc : contact description ; o minus c, reduced time : time) ;
         personel archive
      (nobs : observation number ; oc : time ;
       dc : contact description) ;
    worktab  wt
      (nobs : observation number ;
       dc : contact description ; oc : time)
enddb
```

Fig. 6

```
r in mercury transit ::
     ((same year (r : reduced time, dnt calend (1973))
         and  (r.nc. = dnt contact number (3)))
     and (all t in mercury transit
          ((same year (t. reduced time, dnt calend (1973))
               and (t.nc. = dnt contact number (3)))
          impl  (abs (r.o. minus c)≫ abs (t.o. minus c)))))
```

Fig. 7

```
....

"INTEGER" N ; "REAL" A, B ; ...

B : = 0 ; N : = 0 ;

foreach r in personal archive

    with r. oc. ---> cvt  julian (A) do

        B : = B + A ; N : = N + 1

    enddo ;

"IF" N  ≠ 0 "THEN" B : = B/N ;

...
```

Fig. 8

```
...

"REAL" A, C1, D1 ; "INTEGER" K ;

"REAL" "PROCEDURE" F (X, I) ; ...

update r in personal archive ::
    not  ((cvt julian (C1) ≤ r. oc) and  (r.oc ≤  cvt julian (D1)))

    with  r. oc. <---> cvt  julian (A),

             r. nobs ---> cvt   observation number (K) do

        ... ; A : = F (A, K) ; ...

      enddo ;

...
```

Fig. 9

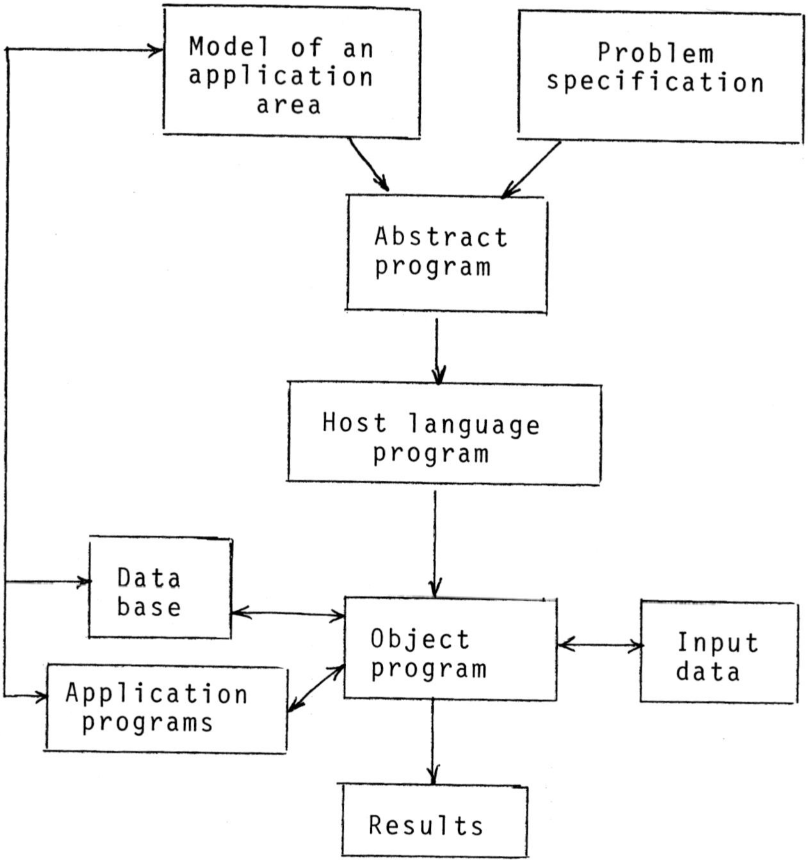
Model of an application area
Problem specification
Abstract program
Host language program
Data base
Object program
Input data
Application programs
Results

Fig. 10

primitive types and maps used in a model must be declared beforehand.

From a model and a problem specification the system synthetizes an abstract program, expressed in terms of abstract types and maps. The system uses here properties of abstract objects, contained in their declarations.

Then the abstract program is translated into a host language. This is performed using properties of physical representations and their programming implementations.

At the next stage the host language program is compiled into an object code. The compilation is done by a serial host language compiler.

The object program accepts input data prepared by the user and returns to him the results of its execution. It may also get information from the data base related to the supporting area model and put some part of its results there. It may also use application programs (modules) from the package and deliver new ones to it.

The sources of the underlying ideas used in SPORA are :

- relational model of data bases (E. F. Codd)
- the language UTOPIST (E. H. Tyugu)
- abstract data types (B. Liskov, E. Moss)
- abstraction levels (E. W. Dijkstra)

and others.

REFERENCE

I.O. Babaiev, F. A. Novikov, T. I. Petrushina. Iazyk Dekart - vhodnoj iazyk sistemy SPORA : - In "Prikladnaia informatika", vyp. 1, 1981, "Finansy i statistika", p.p. 35-73

E. BIBLIOGRAPHICAL SERVICES

PROSPECTS FOR AUTOMATED SOLUTION OF THE SUBJECT CHARACTERIZATION PROBLEM IN THE BIBLIOGRAPHIC SERVICES

S. Schiminovich
American Institute of Physics, 335 East 45 Street,
New York, NY 10017, USA

ABSTRACT.-Evidence is presented on the feasibility of an automated generation of classification schemes and the subsequent indexing of papers with the automatically generated subject headings. The output generated for Astronomy and Astrophysics compares favorably with the schemes used by the services and is not inconsistent with recently proposed classifications. Implementation of the automatic techniques seems not only feasible but desirable. However, it would require that the services include for each paper in their database its citations to other papers.

1. INTRODUCTION

Notwithstanding the continuing search for new techniques in information retrieval and the increasing role of computerization, the intellectual steps that go into the preparation of subject indexes by bibliographic services still follow traditional procedures which are more than half a century old. Computerization has had an impact in the clerical aspects of the task; the construction of a working classification scheme for the subject characterization of scientific journal literature and the indexing process itself are intellectual processes that are still carried out manually.

Mounting costs in the processing of the literature would alone justify looking for a computerized alternative, even without an increase in the overall level of quality performance. In this paper we will show that the present state of the art indeed offers such a computerized alternative; that both, the generation of a subject classification and the assignment of papers to the relevant subject categories are tasks that can be performed automatically at a quite improved level of quality compared with the results of the traditional procedures.

After a brief discussion of the algorithm for automatic classification, we will describe its actual implementation in the generation of the first such output ever generated in the fields of Astronomy and Astrophysics.

C. Jaschek and W. Heintz (eds.), Automated Data Retrieval in Astronomy, 147–154.
Copyright © 1982 by D. Reidel Publishing Company.

We will then examine the output of the algorithm and see that it is not inconsistent with some recent proposals for classification schemes, and could be a useful tool if used as supporting material for their development. The discussion will also help explain some shortcomings of the manual, traditional procedures. Mechanization of the classification procedure is not only technologically feasible, it is desirable.

We will end our presentation with an assessment of the likelihood that a bibliographic service will adopt such computerized techniques and the directions in which research should proceed to improve this likelihood.

2. THE ALGORITHM FOR AUTOMATIC CLASSIFICATION

We have used the BPDA (Bibliographic Pattern Discovery Algorithm), introduced and described a number of years ago in references 1 and 2.

The procedure, graphically represented in fig. 1, is one of successive approximations. At a given step (n) in the procedure, bibliographies $Bj^{(n-1)}$ generated at the previous (n-1)th step act as triggering files to provide input to the algorithm. A link matrix $Lj^{(n)}$ describing the bibliographic links between the triggering file and the Comparison File is constructed, and the BPDA then determines a set of groups $Gi^{(n)}$ and corresponding bibliographies $Bi^{(n)}$ for the (n)th step in the approximate procedure.

The $Gi^{(n)}$ groups are a partitioning or classification of the Comparison File to n-th order in the appoximation procedure, the corresponding bibliographies $Bi^{(n)}$ can act as kernels for operators to retrieve items relevant to the corresponding subject areas from files similar in structure to the Comparison File.

The available database usually acts as the Comparison File, and its structure determines which links are feasible for the construction of the Link Matrix. Up to now the BPDA has been tested on databases in which the inclusion of references to other papers (citations) are a prominent feature of the database (as opposed to abstracts). Therefore, the links are essentially the measures of biliographic coupling originally introduced by M. Kessler[3], and in this sense this work is an extension of his pioneering work.

Few new features were introduced into the original BPDA to increase its efficiency. They rely on a symmetry property of the BPDA:

$$G \rightarrow B, \quad B \rightarrow G, \quad L \rightarrow L^T,$$

where L^T is the transpose of the Link Matrix. This symmetry transformation corresponds in the Comparison File to a "time reversal", a resorting of the Comparison File in which citant takes the place of cited and viceversa.

This inherent symmetry of the clustering procedure over the network of citation links proves that too much has been made of the so called "co-citation coupling". The concept is just the time reversed of the citation coupling originally introduced by Kessler, and rather than adding anything new, reflects the fact that for whatever practical reasons, a transposed Comparison File is being used.

In the present implementation of the BPDA, both Comparison Files, the direct and the transposed, have been constructed and the procedure alternates between them from step to step of the approximation, generating in alternative fashion groups $Gi^{(2n+1)}$ and Bibliographies, $Bi^{(2n)}$.

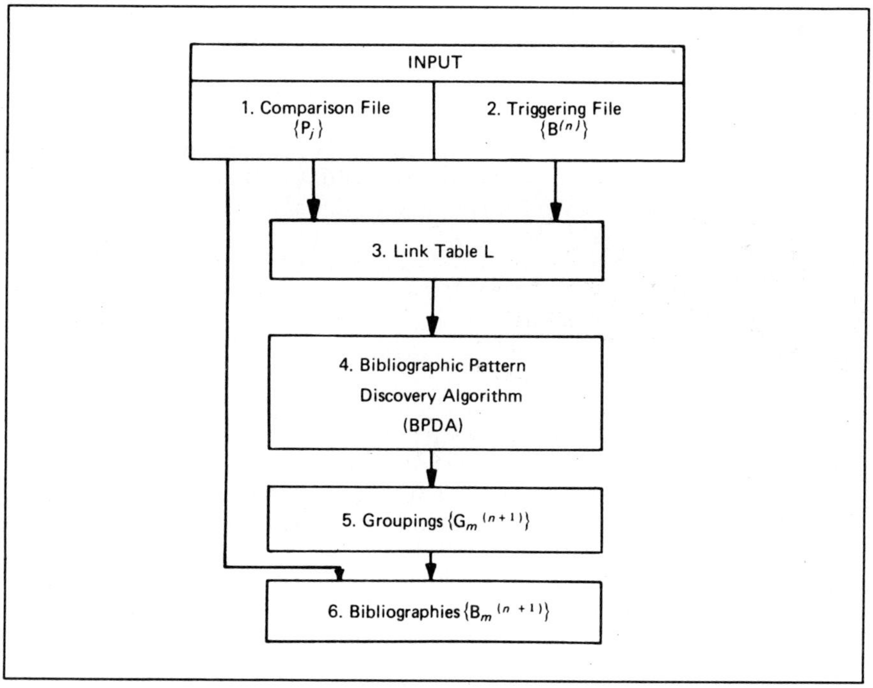

Figure 1.

3. IMPLEMENTATION OF THE ALGORITHM

The results presented in this paper stem from the first full scale computerized implementation of the BPDA. The available database was the complete set of 1978 ISI Citation Tapes, comprising the more than 4,000,000 references contained in about 400,000 papers published that year in different areas of science and technology.

The computer configuration used was a Datapoint micoprocessor with 64K RAM attached to a 100MB storage-capacity disk operating system. The database was loaded on the disk storage in the form of two Comparison Files of about 20MB each and an auxiliary file of about 25MB containing bibliographic information to be used for display purposes.

The typical computation time for performing each of the steps of the approximation procedure described in the previous section was approximately 5 to 10 seconds. Four steps of the approximation procedure proved to be adequate for our purposes. Thus, for a given triggering paper chosen from a representative journal in the Comparison File, there were produced a $G^{(1)}$, a $B^{(2)}$, a $G^{(3)}$, and a $B^{(4)}$; the whole process taking, on the average, half a minute of computation time.

Identification of the subject headings corresponding to each grouping had to be done through manual examination, since no words from either titles or abstracts were available in our database. The task proved quit straight forward and unambiguous, given the close link of relatedness between the papers in a grouping. Since the B's are retrospective bibliographies with frequently cited papers in a grouping, a convenient approach was to examine the papers included with highest weight in a $Bi^{(4)}$. These turn out to be those papers from the past which have been seminal in shaping the area of research in question.

The objection may be raised that the manual determination of the names of the groupings vitiates the very automaticity of the whole process. More important than the feasibility of constructing automatically fair approximations to the name (as shown in Reference 1), is in our opi ion the fact that

1) The groupings generated by the algorithm can be named in such a way that their membership will produce simultaneous high relevance and recall ratios;
2) The same high relevance and recall ratios will be obtained when retrieval for the generated subject headings is continued on other Comparison Files (or data bases) with retrieval operators that use the B's as their kernels.

In other words, the algorithm has uncovered a structure which is intellectually satisfying and which is stable, that is, one which is not sample-dependent.

4. RESULTS FOR THE FIELD OF ASTROPHYSICS

We have used as a triggering file articles in our own database from the journals "Astronomy and Astrophysics", "Solar Physics", and "Astrophysical Journal". Thus far a set of 100 different groupings have been produced and evaluated; the corresponding subject headings being a 15 to 20% sample out of 500 to 800 expected, by extrapolation, for this area of research.

We cannot present here the subject headings, due to a lack of space. To give a flavor of the generated output, however, we will list a dozen, selected from the subfield of Interstellar Matter:

Dynamics of Molecular Clouds.
Molecule Formation in Interstellar Medium.
Molecule Hydrogen in Interstellar Clouds.
Molecular Species in Interstellar Clouds.
Theoretical Identification of Radicals in Interstellar Clouds.
Studies of CO Clouds.
Silicate and Ice Grains in Interstellar Clouds (IR Observations).
Chemical Composition of Hii Regions.
Far Infrared Observations of Hii Regions.
Recombination Lines and Masers in Hi and Hii Regions.
Coronal Gas in the Galaxy (Ovi Absorption Lines).
Supernova Remnants and Galactic Cosmic Rays.

The tenor of these headings, as well as our estimate of their total number shows that the degree of specificity achieved through the automatic procedures goes one level of depth beyond that used by the services in their classification schemes. See, for example, those used by "Astronomy and Astrophysics Abstracts"[4] or sections 95 to 98 of the "ICSU AB International Classification for Physics".[5]

This greater level of specificity is, however, similar to that being proposed for new, revised schemes for Astronomy and Astrophysics as, for example, the joint UDC revision of FID and commission 5 of the International Astronomical Union.[6] Greatest similarity is encountered with a scheme published by VINITI in 1981.[7] Compare, for example, our headings with section 41.25.29, Interstellar Medium (ISM):

Atomic component of ISM; H II regions
Molecular component of ISM; cosmic masers
Dust component of ISM
Compact regions of ISM
Intercloud medium and clouds in the ISM
Gas--dust complexes in ISM
Dynamics and evolution of ISM
Emission sources of unestablished nature
ISM in the solar neighborhood

The possiblity of achieving this greater level of specificity directly, as raw material from the output of the algorithmic procedure, is of no little importance, since it allows us to build classification schemes upwards, instead of downwards on the levels of a hierarchy, as is customarily the case. Our experience with the task of establishing the ICSU AB Classification System for Physics[8] seems to indicate that within the higher levels of the hierarchy, i.e., within the first two levels of that system, there is a better consensus between scientists and specialists on how to construct the classification than within the third and fourth levels. There, a paradoxical lack of consensus arises between scientists as to the proper intellectual characterization of their fields of endeavor and this lack of consensus seems to disappear at the deeper level of specificity encountered by our algorithmic procedures. At this level there is a closer identification with the concrete subject meaning of their work.

Several reasons of a practical nature have kept the services from offering the subject characterization of their databases at the very specific level which scientists would in all likelihood prefer. The less specific levels used do present the material with a vocabulary that eludes the specialized user, a fact that may explain his chronic lack of interest or, sometimes, outright animosity towards the problem of subject characterization of his fields of research.

Differences in the interpretation of subject categories may explain, in part, poor indexing consistency, which is also a problem with the manual systems. The ICSU AB classification, adopted by four major services for the processing of similar material, provides us with an opportunity for conducting large scale studies of indexing consistency between independent teams of indexers. A pilot study that we have conducted shows that two-way agreement for the classification of a document stands at about 35%, and that three-way agreement drops to about 20%. It is then difficult to see how the manual procedures could match the high relevance and recall ratios for algorithmically generated groupings, shown in previous tests to be, simultaneously, both in the vicinity of 90%.

The tardiness with which the manual system is capable of introducing new subject headings to update itself, in response to developments in the field, is another weakness which can be illustrated with an examination of the output of the algorithmic procedures. The concepts uncovered by our analysis were potentially available to automatic systems since 1978, if not earlier; yet, many of them are just filtering into the manual systems. Artificially imposed delays, such as a rule not to update the system more often than a set number of years, should be avoided for proper delivery of a service. Very generally, an algorithmic system would provide for a service a foundation based on objective measures of the biblio graphic properties of the database; manual systems, on the other hand, are much too dependent on the accidents of implementation.

5. PROSPECTS FOR THE FUTURE

Evidence has been given that an automatic option is available to the services for the subject characterization of their databases. We believe that the state of the art makes this option not only possible, but, desirable. However, notwithstanding improvements in the products to be delivered and savings in the operating costs, we believe that the prospects for the adoption of the automatic option are very dim at present. The main obstacle is the requirement to incorporate in their databases the references of each document to other documents, in order to achieve high quality results with the algorithmic techniques.

The problem is not one of economics in the long run, since the input of citations is less expensive than manual classification of a document. The problem is rather one of initial investment and break of inertia to follow customary procedures.

For this reason it would seem worthwhile to pursue further work on algorithmic analysis exploring the possibility of achieving comparable results to those presented in this paper through the use of abstracts instead of citations for the construction of bibliographic links between documents. While preliminary results have not been very encouraging in the past,[9] not all avenues have been exhausted.

In the meantime, we believe that an effort should be mounted to algorithmically map all of science through the use of available citation databases. The massive determination of all the algorithmically possible subject headings, together with the uncovering of interrelationships between corresponding groupings, is interesting and has value of its own, and may become a standard against which to measure further developments in subject classification schemes.

REFERENCES

1) Schiminovich, S.: Automatic classification and retrieval of documents by means of a bibliographic pattern discovery algorithm. Informat. Storage Retrieval, vol. 6; pp. 417-435 (1971)
2) Schiminovich, S.: An Automatic Classification of Bibliographic Data Bases. Biosci. Commun. 1; pp. 24-39 (1975)
3) Kessler, M. M.: Bibliographic coupling between scientific papers. Am. Docum. 14; pp. 10 (1963)
4) "Astronomy and Astrophysics Abstracts", Published for the Astronomisches Rechen-Institut by Springer-Verlag
5) "International Classification for Physics 1977", 2nd ed. (revised and expanded). Paris: International Council of Scientific Unions-Abstracting Board (1977)
6) Draft proposal published by FID in collaboration with Commission 5 of IAU.
7) We are indebted to R. B. Rodman for calling our attention to this scheme published in 1980 in Referativnyi Zhurnal Astronomiya, Geodeziya as well as for providing us with an English translation.

8) Berthelot, A., Clague, P., Schiminovich, S., Zwirner, W.: The ICSU AB International Classification System for Physics: Its History and Future. J. Amer. Soc. Information Science, 30, pp. 344-352 (1979)
9) Feinman, R. D., Kwok, K. L.: Classification of scientific documents by means of self-generated groups employing free language. J. Amer. Soc. Information Science, 24, pp. 382-396 (1973)

THE BIBLIOGRAPHICAL CARD-CATALOGUE OF THE I.A.P. LIBRARY

C. Chagnard-Carpuat, A. Lagorce, S. Laloë
Institut d'Astrophysique
98bis, Bd. Arago, F-75014-Paris

This bibliographical service is available at the Library of the "Institut d'Astrophysique de Paris". It is a hand-operated system, consisting of a card-catalogue concerning all astronomical objects (except those of the solar system). It was started in 1962 and has been updated regularly ever since. It has now more than 200.000 citations for about 19.000 entries arranged in wide sections and, inside each section, in alphabetic order. The main sections are : stars, novae and supernovae, radio-sources, X and γ sources, star clusters, clusters of galaxies, clouds, infrared objects. New sections are added when necessary. Any object is taken, even if it is not clear where to classify it (for example, supernova remnants can be found under supernovae or under X-ray sources).

50 journals are analyzed (including original russian literature) within 2 or 3 days after their arrival at the library. The main limitation is that papers studying more than 20 objects are excluded. These papers are considered as lits and are referenced on special cards : lists of variable stars, lists of double stars, lists of galaxies, lists of H II regions, etc...

Each card gives the complete bibliographical references and, when available, the characteristics of the object (see examples on figs. 1 and 2). To help with nomenclature problems, we use cross-reference cards, as, for instance :

W 80	see	NGC 7000
Sh 117	see	NGC 7000
North America Nebula	see	NGC 7000
W 50	includes	SS 433
SS433	in	W 50

but those cross-references are not given systematically.

C. Jaschek and W. Heintz (eds.), Automated Data Retrieval in Astronomy, 155–158.
Copyright © 1982 by D. Reidel Publishing Company.

A first improvement to the system has just been completed with the use of a word-processing typewriter to print some references.

The next step will be the production on a magnetic disc of the bibliography of non-stellar objects, with the use of the HP 1000 computer of the Institute. A code is already given to each paper (in agreement with the CEDAG, Centre de Données des Amas de Galaxies). For the computerized version, we will not classify the objects in sections but produce only one list in alphabetic order. For example, under the letter "B" :

B... (quasars)
B2...
BDA...
BP...
BRABCMS...
Baade...
Barnard...
Barnard's Loop
Basel...
Becklin-Neugebauer source
Berkeley...
Blanco...
Bochum...
Bootes
Burbidge Chain
Burnhamsche Nebel
Byurakan...

Copies on magnetic tapes will be made and given to the C.D.S. It will then be possible for more users to have access to our catalogue.

At the present time, the bibliography is however already available to anyone interested in a given object : we send copies of our cards on request.

Markarian 35 (1950) α = 10h 42.3 m
δ = 56° 14'

NGC 3353
Haro 3

m_{pg} : 14

WEEDMAN (D.W.) KHACHIKIAN (E.Y.). - Spectral observations of Markarian galaxies with ultraviolet continuum. II. (Astrofizika, 5, 113-122) 1969

WEEDMAN (D.W.). - High velocity gas motions in galactic nuclei (Astrophys. J. 159, 405-413) 1970

WEEDMAN (D.W.). - Emission-line intensities and UBV magnitudes for twenty-three Markarian galaxies (Astrophys. J. 171, 5-12) 1972 & (A.J. Dyer Obs. reprint N° 62)

NEUGEBAUER (G.) et al. - Optical and infrared spectrophotometry of 18 Markarian galaxies (Astrophys. J. 205, 29-43) 1976

ARAKELIAN (M.A.). - The dependence of emission line intensity of Markarian galaxies upon colour index (Astrofizika, 12, 559-562) 1976

HUCHRA (J.P.). - Star formation in blue galaxies (Astrophys. J. 217, 928-939) 1977

ALLOIN (D.) BERGERON (J.) PELAT (D.). - Properties of a sample of irregular galaxies (Astron. Astrophys. 70, 141-149) 1978

Fig. 1. Markarian 35

Cygnus X-1 (1950) α = 19h 56m 28s

HDE 226 868 δ = 35° 03' 55"

BD + 34° 3815

V 1357 Cyg

2U 1956+35

BRINI (D.) et al. - Cosmic X-ray sources in the 20-180 KeV energy range (Astrophys. J. 149, 429-433) 1967

...

...

TREVES (A.) et al. - Ultraviolet X-ray and infrared observations of HDE 226 868 = Cygnus X-1 (Astrophys. J. 242, 1114-1123) 1980 /80APJ02421114/

DANIEL (J.Y.). - The inclination of the orbital plane of Cyg X-1 : a Monte-Carlo study (Astron. Astrophys. 94, 121-125) 1981 /81ANAP0940121/

BALOG (N.I.) et al. - Interpretation of light curves of X-ray binary systems. The system Cyg X-1 (Astron. Zu. 58, 67-79) 1981 /81ASZU0580067/

GOLDMAN (I.). - Cyg X-1 : a massive neutron star ? (Astron. Astrophys. 97, 219-222) 1981 /81ANAP0970219/

KEMP (J.C.) et al. - The 1980 June X-ray high state of Cygnus X-1 : optical changes (Astrophys. J. 244, L73-77) 1981 /81APJ0244L073/

GILES (A.B.). - Observations of sub-millisecond bursts from Cygnus X-1 (Mon. Not. R. Astron. Soc. 195, 721-731) 1981 /81MONN1950721/

Fig. 2. Cygnus X-1 : top of the first card and last card (p.39)

ASTRONOMY AND ASTROPHYSICS ABSTRACTS: A STATUS REPORT

L. D. Schmadel
Astronomisches Rechen-Institut,
Heidelberg, Germany, F. R.

ABSTRACT

The astronomical bibliography 'Astronomy and Astrophysics Abstracts' (AAA) is the direct successor of the well-known 'Astronomischer Jahresbericht' (AJB). AAA is now going into its 13th year of existence. It is the aim of this contribution to outline the present status and the future plans of the abstracting and indexing work at the Rechen-Institut.

INTRODUCTION

The AJB was founded in 1899 by W. F. Wislicenus, the former director of the Strasbourg Observatory. It was the very first astronomical documentation to appear on a periodical basis. Since 1911 this undertaking has been continued by the Rechen-Institut which carried this task until now without any break for 70 years. The AJB was published almost entirely in German and essential parts of the abstracts were produced by external reviewers. Finally, the conventional letter-press printing technique caused some time delay between the receipt of the original material and the record in the AJB. A review of historical aspects of astronomical documentations is given elsewhere (Schmadel, 1979). Fricke (1969) announced the suspension of the publication of the AJB and introduced AAA, for which the following objectives were set:

a) The abstracts will preferably be given in English.
b) Preference will be given to the authors' abstracts with editorial revisions in cases in which a change is considered necessary.
c) The reviewer system of the AJB will be replaced by an editorial staff of the Rechen-Institut.
d) The classification system of AAA will follow the philosophy of the AJB scheme as far as possible.
e) Apart from the author index, an additional subject index will be established. Both indexes are to be produced by a computer in order to serve for retrieval purposes.
f) Two volumes are scheduled to appear per year. The time interval between receipt of documents and the publication should not exceed eight months.

C. Jaschek and W. Heintz (eds.), Automated Data Retrieval in Astronomy, 159–161.
Copyright © 1982 by D. Reidel Publishing Company.

g) The letter-press technique will be replaced by the offset reproduction method. The text will be written at the Rechen-Institut composers.

This fundamental concept is still valid. The most important change will be the introduction of a computer-aided text processing system.

THE ACTUAL WORK AT AAA

From the receipt of a document until the delivery of the composer copy there are three main processes:

a) classification and indexing,
b) text recording and corrections,
c) manuscript and index compilation.

Scientific treatment, classification. As a rule, we only use the original material. The bibliographic remarks and all further annotations are done by pencil. Concerning the main subject of the paper, a classification number is given. In many cases this obviously is not sufficient to describe all aspects of the paper. Then, a so-called cross reference containing the title of the paper and a reference to the main category is given. The abstract consists of either the authors' text or an abridgement or modification or a completely new version. The classification step ends with the declaration of some key words, the abbreviation of the source, and – if necessary – the transliteration of Slavic author names. A running number within each subject category will be added.
Text recording, corrections. The technical staff now records the complete bibliography consisting of the title, the author names, the source of information, the number of the main category, the running number, and an abstract. This is actually done by means of IBM 72 composers that have the capacity to type some 800 different characters. Two corrections are made with the aid of a jointly produced duplicate of the abstract strip. The index informations are recorded on punched cards followed by two correction and culling steps as well.
Manuscript and index compilation. The editorial closing date lies some two months after the end of the preceding half-year period. After this date all abstract and cross reference strips are sorted, compiled into complete pages, and are corrected once more. Chapters with cumulative informations can be compiled only at this stage. The index informations are treated by our computer. The line printer output is the direct copy for the offset reproduction. The compilation process of the whole manuscript takes another two months of work.
Statistics. We examine each item of our volumes at least five times. The achieved error rate per printed character is of the order of a few ppm. The AAA staff actually consists of only five scientists, one translator for the Russian publications, and four secretaries. We are supported part-time by three assistants. The amount of material is rather large in relation to these numbers. Each volume contains some 8,500 references to documents of almost 10,000 authors. The list of periodicals scanned consists of some 700 titles, and this will be enlarged by some 100 publications of observatories and institutes. Since 1969, we have recorded more than 170,000 documents of all kinds. The total number of fully retrievable punched card informations amounts to some 540,000 items. It is due to these numbers that we are forced to produce general index volumes of accumulated key words and author names after the publication of every ten ordinary volumes.

FURTHER ACTIVITIES AND FUTURE PLANS

Actually, we use more than 20,000 different key word combinations. These so-called free descriptors will provide a possibility to perform retrospective literature searches. At present we work on a condensed version of this vocabulary of astronomical terms. This collection could serve as a first step towards the construction of an astronomical thesaurus.
In the preparation process of the manuscript there are many steps which could be avoided by using a computer. The text recording should be done by intelligent terminals. The procedure will yield a complete pattern ready for photocopy printing. As a by-product, all index informations are available immediately. We hope that a large amount of technical work could be drastically reduced. The use of electronic means also gives the possibility to form a data base. We do not intend to publish a magnetic tape service of AAA, but we plan to incorporate our material within the INKA-PHYS data base of the Karlsruhe 'Fachinformationszentrum' which already provides literature searches from all fields of physics. First considerations concerning the necessary software developments have shown that the overall organization of our service has to be changed only slightly. This somewhat revolutionary adaption within the documentation history of the Rechen-Institut should be attainable in the next few years.

REFERENCES

Fricke, W. 1969, *Mitt. Astron. Ges.*, No. 27, pp. 125 - 126.
Schmadel, L. D. 1979, *Bull. Inform. CDS*, No. 17, pp. 2 - 11.

INSPEC ACTIVITIES IN ASTRONOMY AND ASTROPHYSICS

Mike J. Collins
INSPEC, The Institution of Electrical Engineers,
Hitchin, England

Considers a typical paper from the astronomical literature, shows how it was processed for 'Physics Abstracts' and illustrates how it can now be retrieved on-line from the INSPEC database.

"Two-color photoelectric observations of the eclipsing binary BB Peg" by Cerruti-Sola and Scaltriti was published in Astronomy & Astrophysics Supplement Series, Vol. 40, No. 1, p. 85-9 (April 1980). The abstract for this paper appeared as No. 62194 in Physics Abstracts, Vol. 83, No. 1139, p. 4806 (1 July 1980). This particular journal article has been chosen to illustrate the way material is handled by the INSPEC database because apart from the usual bibliographic elements, a designated stellar object (BB Pegasi) is mentioned and numerical data (observation dates) are incorporated.

The editorial work was performed on two sheets: the title page TORN from the journal (Figure 1a) and an SA2 'worksheet' (Figure 1b). This paper was the seventh in a single batch of 11 papers and the batch code (A351-8004-L) can be seen on the top left of the title page showing that this was the fourth issue received in 1980 from the journal Astron. & Astrophys. Suppl. Ser. (A351). A label was attached to the title page indicating the paper number (A7), the pagination and the number of references given on page 86 of the article. Details on the title page include capitalization in the title (by triple underlining), editing the original abstract which in this case resulted in a shortening by three lines, and selection of uncontrolled (or "free") index terms which were underlined in the title or abstract.

The SA2 can record the language, classification codes, a common 'modifier line', controlled index headings with their modifier lines, bibliography topic title (for papers with references >50), additional free indexing, and treatment codes. An overflow sheet is available. Labour-saving devices include a common modifier recalled by using 'ZZ' (ZZ Ceti stars cause problems!) and abbreviated index headings (VISAO generates the heading "visible astronomical observations"). The 'CC'

C. Jaschek and W. Heintz (eds.), Automated Data Retrieval in Astronomy, 163–165.
Copyright © 1982 by D. Reidel Publishing Company.

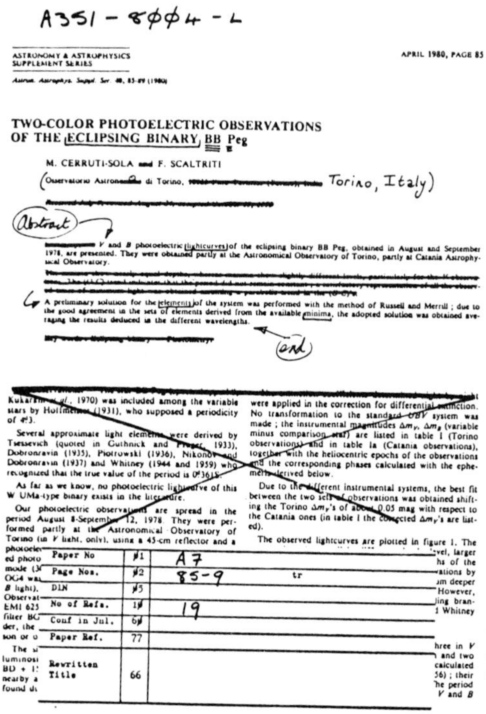
A351 - 8004 - L

ASTRONOMY & ASTROPHYSICS SUPPLEMENT SERIES

APRIL 1980, PAGE 85

Astron. Astrophys. Suppl. Ser. 40, 85-89 (1980)

TWO-COLOR PHOTOELECTRIC OBSERVATIONS OF THE ECLIPSING BINARY BB Peg

M. CERRUTI-SOLA and F. SCALTRITI

(Osservatorio Astronomico di Torino, Torino, Italy)

Abstract

V and B photoelectric lightcurves of the eclipsing binary BB Peg, obtained in August and September 1978, are presented. They were obtained partly at the Astronomical Observatory of Torino, partly at Catania Astrophysical Observatory.

A preliminary solution for the elements of the system was performed with the method of Russell and Merrill; due to the good agreement in the sets of elements derived from the available minima, the adopted solution was obtained averaging the results deduced in the different wavelengths.

Paper No	01	A7
Page Nos.	02	85-9
DIN	05	
No of Refs.	10	19
Conf in Jnl.	60	
Paper Ref.	77	
Rewritten Title	66	

Fig. 1a Edited title page

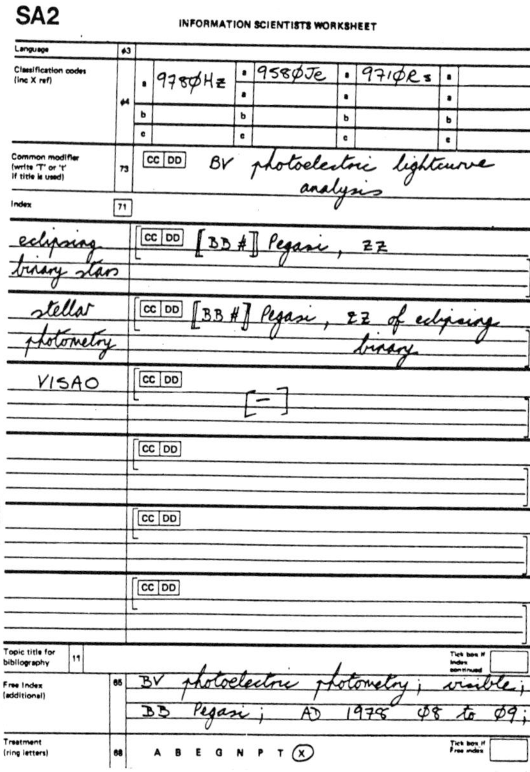
SA2 INFORMATION SCIENTISTS WORKSHEET

Language	03	
Classification codes (inc X ref)	04	9780Hz; 9580Je; 9710Rs
Common modifier (write 'T' or 't' if title is used)	73	CC DD BV photoelectric lightcurve analysis
Index	71	
eclipsing binary stars		CC DD [BB #] Pegasi, ZZ
stellar photometry		CC DD [BB #] Pegasi, ZZ of eclipsing binary
VISAO		CC DD [-]
		CC DD
		CC DD
		CC DD
Topic title for bibliography	11	
Free Index (additional)	65	BV photoelectric photometry; visible; BB Pegasi; AD 1978 08 to 09;
Treatment (ring letters)	68	A B E G N P T (X)

Fig. 1b SA2 worksheet

box is used for chemical sorting and 'DD' for elementary particles. The [[ ]] facility in this case, takes BB out of sort so that all the Pegasean objects are kept together. Our twice-yearly subject indexes of four-yearly accumulative indexes will reveal the result of sorting control.

Only the modifier lines to the fixed index headings are lost to on-line retrieval because they are used solely for the printed indexes. All the other fields can be retrieved on-line and a simplified example illustrating the various ways that this paper could have been found using the Lockheed DIALOG service is shown. A copy of this retrieval example with full explanation of the search strategy and responses can be supplied free upon request. Those of you familiar with DIALOG will only need to have sets 13-18 explained. Set 13 selects any papers describing observations starting or ending in 1978 August, sets 14-15 plus set 17 select papers beginning in 1978 and ending in August of that year. Set 18, which is a combination of sets 13 to 17, therefore presents papers describing observations commencing or ending in 1978 August. The exhaustive case is possible but more complicated!

I would like to thank Professor Lequeux for permission to use an Astron. & Astrophys. Suppl. Ser. paper as an example.

```
B13
FILE 13: INSPEC - 78-81/ISS12
SEE FILE 12(1969 THRU 1977)
          SET  ITEMS DESCRIPTION

? S AU=CERRUTI-SOLA?(C)AU=SCALTRITI?
            1       2 AU=CERRUTI-SOLA?(C)AU=SCALTRITI?
? S CC=A9780H
            2     775 CC=A9780H
? S CO=AAESB9
            3     530 CO=AAESB9
? L3/ENG
            4     462 3/ENG
? S PY=1980
            5  135011 PY=1980
? S CS=TORINO OR CS=TURIN
                  853 CS=TORINO
                  110 CS=TURIN
            6     933 CS=TORINO OR CS=TURIN
? S BB(W)PEG?/TI
            7       2 BB(W)PEG?/TI
? S ECLIPSING BINARY STARS/DE
            8     543 ECLIPSING BINARY STARS/DE
? S LIGHTCURVE?/ID
            9      63 LIGHTCURVE?/ID
? S PHOTOELECTRIC?(F)LIGHTCURVE?
           10      46 PHOTOELECTRIC?(F)LIGHTCURVE?
? S ECLIPSING(W)BINAR?(C)LIGHTCURVE?
           11      34 ECLIPSING(W)BINAR?(C)LIGHTCURVE?
? C 1-11/AND
           12       1  1-11/AND
? S AD(4W)1978(W)08
           13      73 AD(4W)1978(W)08
? S AD(W)1978(3W)08
           14     102 AD(W)1978(3W)08
? S AD(W)1978(1W)08
           15      78 AD(W)1978(1W)08
? C 14-15
           16      24  14-15
? S AD(W)1978(1W)08(1W)08
           17      78 AD(W)1978(1W)08(1W)08
? C 13OR16OR17
           18     121  13OR16OR17
? C 12AND18
           19       1  12AND18
```

Fig. 2 Search example on DIALOG showing how various characteristics of the Cerruti-Sola and Scaltriti paper on BB Pegasi could have been used for on-line retrieval.

STELLAR BIBLIOGRAPHY RETRIEVING SYSTEM IN JAPAN

Shiro Nishimura
Tokyo Astronomical Observatory, University of Tokyo

ABSTRACT

On-line retrieving systems of BSI in Japan is briefly described. The BSI reference file was adapted to be retrieved by some keywords other than star names. Then the retrieval system of BSI was constructed on a general retrieval language CIR. Some efforts to make the system efficient and to save the disk space are discussed.

RETRIEVING SYSTEMS BASED ON A DEDICATED PROGRAM

Since 1978 two on-line retrieving systems of BSI have been operated in Japan. Two systems are almost identical, and the one is on UNIVAC 1100 of the Tokyo Astronomical Observatory, the other on HITACHI M200-H of the Computer Center, University of Tokyo. The latter system can be accessed through the public telephone network by astronomers and related scientists in allover Japan, while the former system is used locally and for the system maintenance. The retrieving program including code and decode subprograms of star names was written in PL/I language. In addition to the BSI reference file and the inverted file of keywords (star names), a third file of the thesaurus of star names was compiled at the Tokyo Astronomical Observatory to make sure hitting of references from variant star names. Data of the file were taken from CSI and other supplementary data. Details of the systems were given elsewhere (Nishimura 1981).

So far BSI can be retrieved only by star names. The usage of delimiters and the abbreviation of journal names were inhomogeneous through the original BSI reference file. Furthermore the program was so simple that almost no logical operations could be applied to the sets of the retrieved references. In order to make the retrieving system more versatile, it will be more profitable to resort to a generalized retrieving language than to develop a complicate unique program. A retrieving language CIR (Conversational Information Retrieval) of UNIVAC Japan

C. Jaschek and W. Heintz (eds.), Automated Data Retrieval in Astronomy, 167–169.
Copyright © 1982 by D. Reidel Publishing Company.

was recently made available at the Tokyo Astronomical Observatory. Although CIR has rather limited functions, it was adopted to make the first trial.

ADAPTATION OF BSI TO DATABASE APPROACH

It was necessary to convert the BSI reference file into a unified form. Processing of the file on the computer was divided into many small steps, and was executed semi-automatically by manual interventions. The final form of each reference was " Author(s) < Journal name > < Volume, (Number,) Page > Title ". (In case of books it was modified as " Author(s) <%T Book title, (%E Editor(s),) (%P Publisher,) (%C Comments) > < Page > (Chapter title) ".) The author entry had a form " Family name, I. I'. ... (, Jr. or III etc.) / ..." where I and I' denote the initials of the first and middle names. The journal names were unified in accordance with the abbreviations in the Astronomy and Astrophysics Abstracts, reducing the number of journal names from 465 to 133. Similarly the number of authors was reduced from about 7000 to 6000.

RETRIEVING SYSTEM BESED ON CIR

The items of the extended keywords are as follow :--- 1. Reference numbers, 2. Star names, 3. Author names, 4. Journal names, 5. Book titles, 6. Titles. The last two items were not registered in the inverted file, and will be retrieved only through the reference sets extracted by other items. All items except star names were derived from the BSI reference file, while star names were acquired from the BSI star file (the inverted file) by re-inversion. As there was no thesaurus function in the CIR system, one could have listed up all the variant star names, which corresponded to each object appearing in a reference, and included them together with the BSI star names as the keywords of the reference. On the average an object has four names. Brighter stars, however, which have more variant names than the average, tend to appear more frequently in the bibliographies. The total number of keywords of star names reached three million, and the maximum number of star names corresponding to a single reference was eighteen thousand, which far exceeded the limitation of CIR. After all only one name of each object, that is the BSI star name, was registered as the keyword.

The CIR system affords the "Library File", which enable the macro-command of retrieval. If a library element named by an variant star name have been created and its contents dictate the retrieval with the keyword of the BSI star name, then one can invoke the retrieval of references relevant to the object simply by entering the variant name. Some star names, such as those in double star catalogues, denote many different objects in BSI. Even in such a case, the summation of the sets of references for each object will be obtained, as the library element may contain several retrieval commands including the sum operation.

Data of the thesaurus file were converted to the elements of the "Library File" in the CIR system. A shortcoming of the Library File was a bulky space in the disk necessary to the on-line retrieval. Star names in some catalogues, which will not often used to denote an object, were deleted from the file. In spite of these efforts, the present size of the library file was 160 MB in addition to 15 MB of the fundamental data files of BSI-CIR system, compared to 8 MB of the total data space in the dedicated program system. The space of the library file will be soon reduced to a half of the present size by the revision of the CIR system program. It is expected to be furthermore compressed to about 20 MB by introducing the variable length structure, since the present file is occupied by meaningful data very sparsely. Then there would be little difficulty in the disk space to make the steady on-line retrieval feasible.

The test of retrieval proved successful. Retrieval by various kinds of keywords could be executed, and the sum, difference or product of sets of references could be obtained in an arbitrary mode. It became easier to reach the reference one wanted to have. The CIR system provides convienient tools of database maintenance. The revision of BSI original file could be performed more easily by reorganization of updated CIR database.

FUTURE PLAN

Another project is now being planned to construct a system based on ORION of HITACHI at the Computer Center, University of Tokyo. It will serve for the nation-wide demands, and the allowance of the disk space will be favorably given to such a commonly used database. If the files are too large to be located in the disk, they will be stored permanently in the "Mass Storage System" and transferred into the disk when used.

SUMMARY

A retrieval system based on a general purpose program was tested. It was revealed that this approach was successful in the retrieval of references in flexible mode of operation, and as the tool of the maintenance of BSI files. The difficulty to hold a large space in the disk will be overcome in near future.

REFERENCE

Nishimura, S. : Bull. Inf. CDS, N. 20, Pp82-85.

THE BIBLIOGRAPHICAL STAR INDEX

F. Ochsenbein
Centre de Données Stellaires (C.D.S.)
Strasbourg - France

INTRODUCTION

The Bibliographical Star Index (B.S.I.) ries to solve the problem of retrieving the observations or any kind of notes about a star. This file was initiated by Prof. R. Cayrel in 1964 and is going on through a collaboration between Paris-Meudon Observatories and the Stellar Data Center (see Ochsenbein and Spite, 1977). In pratice, the astrophysical papers published on stars are read, the names of all the stars cited in each paper are recorded ; simultaneously, the complete bibliographical reference is recorded : name of the author(s), journal, volume, page, year of publication and the full title. These data are keypunched and stored in the computer, each star name is then tested, frequently replaced by a more used designation, and entered into the B.S.I. The retrieval program than lists all the papers with authors, references and full title in which a given star is cited.

DIFFICULTIES AND LIMITATIONS

Many difficulties prevent from reaching the desired completeness of this bibliographical survey.

1.1. The file begins in the year 1950 ; we do not feel, at the present time, that an extension to earlier publications is a strong need.

1.2. Only 12 journals were covered for the years 1950-1972, and about 40 have been covered since, which represent likely 90 or 95 % of the astronomical literature.

1.3. Lists of more than 1 000 objects are excluded ; 35 such lists were found for the period 1950-1978, but it should be noted that the majority of these lists are included in catalogs available on magnetic tape or microfiche at the Stellar Data Center.

1.4. The inextricable system in use for stellar designations (see e.g.

C. Jaschek and W. Heintz (eds.), Automated Data Retrieval in Astronomy, 171–173.
Copyright © 1982 by D. Reidel Publishing Company.

Jaschek et al. 1980) is likely the main difficulty in this enterprise. At the present time, about 200 catalogs with their numbering system are recorded and the growth rate is about 30 new catalogs used for stellar designations per year. A software able to handle this large diversity of stellar designations was created, and the use of the Catalogue of Stellar Identifications (see e.g. Ochsenbein et al, 1981) improves the completeness of the B.S.I., at least for the stars brigther than the 10th magnitude. For the fainter stars, we hope that the situation will be much improved in the near future, with the extension of the C.S.I. to fainter stars which has been undertaken by M. Bischoff.

1.5. In fact, the B.S.I. is not only incomplete in several ways, but it is not error-free. As far as we may state some general conclusions about the discovered errors, our experience is that keypunching errors are very rare compared to other sources of errors - and the confusing system used in stellar designations is likely the major source of errors. We thank here the users who have been kind enough to send us lists of corrections, and we encourage all users to do likewise. Remarks, criticisms and suggestions are welcome.

STATISTICAL SUMMARY OF THE B.S.I.

We summarize in Fig. 1 the contents of the B.S.I. for the period 1950-1978 : an amount of 12 000 papers includes 350 000 citations about individual stars, referring to 89 000 different stars. This results in a mean number of 4 citations per star for this period, but this mean value includes of 61 % stars quoted only once ; Fig. 2 is an histogram of the number of citations per star up to 30 citations ; 1 662 stars with more than 30 citations are not plotted in this figure.

Another precision is the part of faint stars in the B.S.I. : roughly 50 % of the stars are fainter than the HD limiting magnitude but in 1978 HD stars still represent 71 % of the citations ; bright stars are obviously more frequently studied than fainter objects.

AVAILABILITE OF THE B.S.I.

Several facilities have been developed to access the B.S.I. :

a) 3 sets of microfiches covering the periods 1950/1979, 1973/1977 and 1978 are available at the C.D.S.

b) Bibliography about a list of stars is sent on request (print-out form)

c) The B.S.I. can now accessed through TRANSPAC or EURONET networks in an interactive made.

We hope that this last facility will be frequently used in the future -

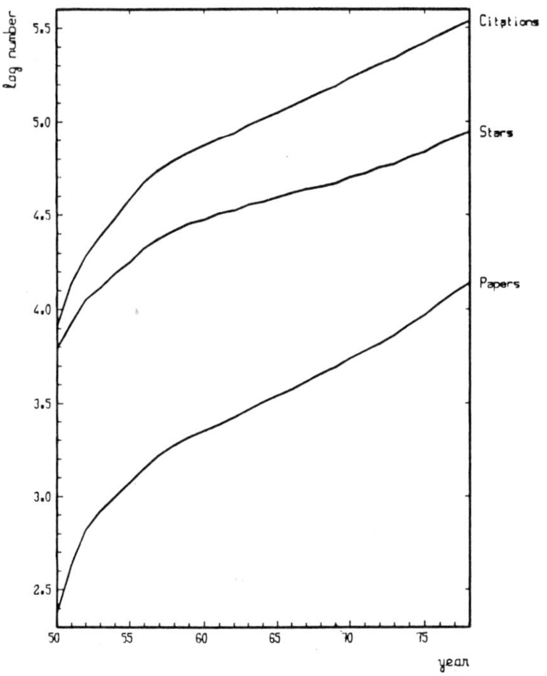

Figure 1

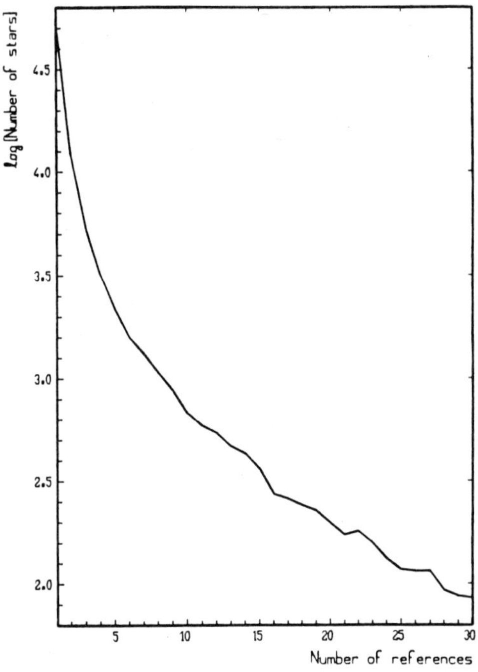

Figure 2

the astronomer gets immediately the more wanted information, and this method implies a reduced amount of work for the staff of the C.D.S.

ACKNOWLEDGEMENT

It is a pleasure to thank the people who are involved in this rather tedious work : M. Bischoff, R. Cayrel, C. Jaschek, S. Kirchner, R. Lahmek, F. Spite, M.J. Wagner and A. Acker and J. Marcout for the part of the B.S.I. dealing with planetary nebulae (Acker et al, 1981).

REFERENCES

Acker A., Marcout J., Ochsenbein F. : 1981, Astron. Astrophys. Supp. 43, 265

Jaschek C., Ochsenbein F., Bidelman W.P. : 1980, C.D.S. Inf. Bull. 18, 41

Ochsenbein F., Spite F. : 1977, I.A.U. Coll. 35, Astrophys. Space Sci. Library, 64, 175

Ochsenbein F., Bischoff M., Egret D. : 1981, Astron. Astrophys. Supp. 43, 259

F. COPYRIGHT

COPYRIGHT AND USE OF COMPUTERS FOR ACCESS TO PROTECTED WORKS

Evgueni Guerassimov
Legal Officer, Copyright Division, Unesco

Copyright

It is considered that the term "copyright" originated in Anglo-Saxon law at the beginning of the eighteenth century. Originally, it meant a "right to make a copy" of a work. This was because at that time copying (mainly by printing presses) was almost the only mode of reproduction.

Despite the evolution of technological means of use of works, this term remained the same though actually it has a dual meaning: the term "copyright" may be interpreted as an institution of law and as rights vested in the creator of a work (author) or other successor-in-title. Compared with analogous terms in other languages, the term "copyright" means, for example, "droit d'auteur" in French, "il diritto di autore" in Italian, "avtorskoe pravo" in Russian, "urheberrecht" in German. These terms seem to be more adequate nowadays as they may cover different means of use of works.

Unlike inventions, where the right is granted upon its recognition by the State as certified by an instrument of protection (patent, etc.), copyright - as a general rule - arises from the very fact of creating a work if it fulfils the requirements of the copyright law. (Only in a small number of States recognition of copyright is subjected to compliance with some formalities such as registration and deposit of the work.)

Copyright law generally covers all kinds of works of literary, scientific and artistic nature, whatever the mode or form of their expression. These may include books and other writings, lectures and works of the same nature, dramatic, dramatico-musical, musical, choreographic, cinematographic, photographic works, works of fine and applied arts, illustrations, maps, sketches and three-dimensional works relative to geography, topography, architecture or science. Derivative works such as translations, adaptations, arrangements of music and other modifications of literary, scientific or artistic works are

C. Jaschek and W. Heintz (eds.), Automated Data Retrieval in Astronomy, 177–184.
Copyright © 1982 by D. Reidel Publishing Company.

usually protected as original works without prejudice to the copyright in the original work. Collections of various works, such as encyclopedias and anthologies which, by reason of the selection or arrangement of their content constitute intellectual works, also enjoy protection. Copyright law as a rule, excludes from its scope official texts of a legislative, administrative or judicial nature, such as laws, decrees, administrative orders, public documents, court decisions, etc. (The reason for this is evident - the content of such texts should reach the general public as rapidly as possible.)

It should be noted that the copyright law does not protect the ideas contained in a work but the form in which they are expressed. This may be illustrated by the following example. A scientist may make a discovery and the same discovery be made by another scientist because both of them are investigating the laws of the objective world. The scientist who manages to patent the discovery first will enjoy the right in it under the law on patents. However, nothing prevents other scientists from writing books dedicated to the discovery in question. Each of these books may enjoy protection under the copyright law because it is excluded that the same subject be described by different authors in exactly the same form (arrangement of the content, the order of words, etc.) unless, of course, it is simply a matter of copying. This applies also to painting, photographing or making a cinematographic work on the same subject. In short, each literary, scientific or artistic work expressed in original form falls within the scope of the copyright law in spite of the fact that it is based on the same idea, subject, historical event, etc.

The law initially vests the copyright in a work in its creator, i.e. the author, because the work represents the fruit of his intellectual labour. There may be cases, however, where the original copyright belongs to third parties. This depends on legal traditions of a country. For example, in the countries with Anglo-Saxon legal approach, copyright in commissioned works or works created under contract of employment belongs to the commissioner or employer unless the contrary is established by the contract of commission or employment. However, in the countries following Roman legal traditions copyright in such cases vests originally in the author except where it was agreed in writing to the contrary. The same relates to copyright ownership in cinematographic works, periodicals, reviews and newspapers.

Copyright comprises two kinds of rights : moral (or non-pecuniary) rights and economic (pecuniary or exploitation) rights.

Moral rights are mainly the right to the authorship in the work, in other words, the author's right to claim recognition of his quality as author of his work, and the right to the integrity of the work, which may be described as the author's prerogative to oppose any distortion, mutilation of, or other derogatory action in relation to his work. Some national laws on copyright also expressly mention the author's right to make corrections in his work before its publication or when a new edition is prepared and the right to withdraw the work from sale or use (upon identifying any interested parties) where it no longer reflects his points of view, convictions etc.

Moral rights are attached to the persona of the author and may not be transferred to third parties. After the author's death they may be exercised by his heirs or other persons designated in his will or by a competent authority designated for this purpose in virtue of law, irrespective of ownership of the economic rights.

The author's economic rights are connected with various means of using a work. Some national laws provide for these rights in a very general form by stating that the author will have the right to control the exploitation of his work. But usually the copyright laws state either the economic rights or the forms of use for which the author's authorization is required. This relates to reproduction of the work by various means; public performance, broadcasting, making of translations, adaptations etc. and other acts by which the work may be communicated to the general public.

Economic rights may be licensed, transferred or assigned by the author inter vivos to third parties against payment or gratuitously, and transmitted mortis causa to his heirs or other persons in accordance with the author's will or in virtue of law. The author may also entrust administration of these rights to his professional organization which will be empowered to act as agent for the issue of authorization and for collection of royalties deriving therefrom.

It should be mentioned that the author's economic rights in respect of his work are not absolute either in their extent or in duration.

The copyright law usually provides for certain limited cases of free use of published works in the interests of society as a whole

For example, any person may make a reproduction or translation of a work for his private use. To some extent, an author may make quotations from works of others to illustrate the content of his own work, criticize others, enter into polemics etc. To the extent justified by the purpose, a work may be used by way of illustration in publications, etc. intended for teaching. Under certain conditions a work may be included in reports on current events by means of photography, cinematography, broadcasting, etc.; periodicals and broadcasting organizations may reproduce or broadcast articles on current, economic, political or religious topics, etc.

There may also be cases where a given work may be used without the copyright owner's consent but subject to remuneration for such use (the cases of compulsory or legal licences). This quite often relates to mechanical reproduction or works (discs, tapes) once the copyright owner has authorized this act for commercial purposes.

Under the national laws of many countries the moral rights are protected perpetually. This, in our view, is aimed at avoiding the danger of distortion and misrepresentation to which works of the mind are exposed. Certain national laws provide for the same duration of protection of moral rights as for economic ones. Economic rights are generally protected from 25 to 50 years after the author's death, (or after the date of publication in case of anonymous and pseudonymous works) depending on national legislation. After expiration of the term of protection, the work falls into the public domain and may be freely used, though some laws require that a small fee be payed.

Development of technological means by which protected works may be used has greatly influenced the evolution of national legislation on copyright. Elementary printing presses were perfected and followed by rotaprint, photomechanical reproduction and microfilming. Photography was followed by the invention of cinematography; radio broadcasting was followed by television. All these inventions enabled wide use and dissemination of protected works both nationally and internationally and required changes in the copyright law in order to maintain the author's control over the use of his work. However, this prerogative of the author is further complicated by the rapid scientific and technological progress witnessed throughout the 1970s. The use of satellites for broadcasting relay, the making of audiovisual recordings, distribution of broadcasts by cable television systems, use of computers for access to protected works, may have the effect of removing the copyright from the author. This is the problem raised by modern technological development for national legislatures and such intergovernmental bodies as Unesco which is administering the Universal Copyright Convention. In this respect it should be underlined that copyright law not only establishes social justice regarding authors but also stimulates creative activity by allowing intellectual persons to live on the income from the use of their works, favouring creation of a class of professional authors. Thus, the copyright law, in the long run, contributes to development of culture in general.

Copyright problems arising from the use of computers for access to protected works (1)

Referring to the latest developments in the field of computers in general, an extract from an article by Brigitte d'Aranda, freelance journalist from Strasbourg may be cited:

"The era of the new technological revolution, a greater revolution than that which replaced human physical effort by the steam engine, is upon us. It is overturning the world economy. Today, a microprocessor a few millimeters thick contains the same processing power as a computer that filled an entire room ten years ago. In five years, its price has shrunk twenty times. Everywhere and in the most disparate fields, from telephone network to factory and including games and teaching machines, micro-electronics have entered the scene and are dominating it." (2)

Computers are increasingly dominating the information scene. Creation of carefully directed information and documentation systems has become necessary for the performance of the various functions of society. Rapid development of computer technology has led to the creation of computerized information systems, networks and data bases both on national and international levels. This enables information-seeking users to have direct access to the information and documentation stored in such systems.

Copyright was not involved at the time when the materials which were stored in and retrieved from the computers represented unprotected matters, such as mere bibliographic data (stored for the purpose of cataloguing) or the texts of laws, court decisions and other materials of the same nature retained in the memory of the computer. But the scene changed when abstracts were included in the bibliographical data and the legal texts were accompanied by commentaries taken from literature, in other words, when copyrighted materials were fed into computers. Nowadays, the information and documentation activities include collection, storage and making available different kinds of literature on a full-text basis. Modern information technology (electronic data processing, microfilm storage, etc.) makes full texts of protected works even more accessible, not only to industrial enterprises, research establishments, public authorities etc., but also to the general public through the libraries. It will not be long before any information-seeking user or person, eg. in France, will have access to the works published and located in the United States by simple manipulation of the computer buttons. Nevertheless, the author or other copyright owner cannot remain indifferent to such unrestricted use of his work. Consequently, the question arises as to whether copyright should come into play.

This question remained in the agenda of the Intergovernmental Copyright Committee of the Universal Copyright Convention and the Executive Committee of the Berne Union for more than ten years. First of all, it was carefully studied (3) and the results of these studies were examined by the Committees. In 1979, a Working Group of Experts from various countries, convened jointly by Unesco and WIPO, drew up important conclusions on the matter. In 1980 the copyright problems posed by computers and the conclusions of the Working Group were discussed by a Committee of Governmental Experts. The Committee generally approved the conclusions reached by the Working Group and further developed and specified them.

The substance of the conclusions reached by the above Working Group and Committee may be summarized as follows:

1. In general, the use of computers for storage and retrieval of protected works does not create a new situation with regard to copyright protection. The existing principles in international conventions and domestic laws on copyright may be applied to the problems raised by this new technique. The question is how to apply these existing principles in order to cover adequately all situations arising from the use of protected works by computer technology.

2. As far as subject matter of storage and retrieval is concerned, three main categories and legal situations should be clearly distinguished:

a) The usual particulars of protected works (name of author, title, publisher, year of publication etc. - so-called 'index method'). Their use as such, for the purpose in question, does not give rise to copyright.

b) Full text of protected works (such as articles, journals and magazines, newspapers, works published in book form, including protected compilations). It is evident that, in so far as a given work is protected under national legislation on copyright, its use by computerized systems should require prior authorization of the author or other copyright owner.

c) Abstracts. As the case may be, an abstract of a work may or may not enjoy copyright protection. Originality and creativity constituting the basic elements justifying copyright protection of any work should also apply to abstracts. The Committee distinguished three main categories of abstracts:

(i) an abstract which is an adaptation of the original work which should be protected;

(ii) an abstract of a pre-existing work (not constituting adaptation) which is original in character and results from the creative efforts of its author, should also be protected;

(iii) an abstract which is limited to a mere enumeration of ideas and facts expressed in the original work does not seem to form a proper subject matter for protection.

Consequently, the use of the first two categories of abstracts should require prior authorization of the author or other copyright owner.

The Committee also recalled the case where abstracts are written by third persons. If an abstract constitutes adaptation or other derivation of a pre-existing work, its writing requires authorization of the author of the full text.

As far as compilation of information stored in an automated data base was concerned, the Committee's opinion was that such compilation should be considered as protected if it is original and results from creative efforts (collection and structuring of the data) and subject to the general rules governing the protection of the relevant category of works provided for in domestic laws. In other words, a compilation resulting from computerization of information may not enjoy copyright protection if it consists of nothing more than added data and no specific criteria or classification method was used for arrangement of its content. To enjoy protection a compilation must reflect the intellectual efforts of its maker. It may be illustrated by the following example. Supposing somebody is preparing a compilation of bibliographical data on publications in the field of medicine. The chosen subject is 'treatment of disease X'. All publications on this subject are collected and classified depending on the methods of treatment proposed or described by different authors. Furthermore, the data is arranged in such a manner as to show the evolution of the various methods and the country of publication. Thus, making of the compilation requires professional knowledge in order to choose the subject, collect and classify the data and arrange its content. The intellectual efforts and personal contribution of the maker seem to be evident, the compilation should be considered as an original one and enjoy protection under copyright legislation. The author of the compilation should be the person who established it before it was fed into the computer. This is because the person who actually introduces the data into the memory of the computer did not participate in the creation of the compilation as such. Admittedly, the questions of granting protection, determination of authorship (initial author or employer, physical person or legal entity, joint or independent authorship, etc.) and the term of protection of this category of works are governed by national laws on copyright.

3. With regard to acts involved in storage and retrieval of protected works, the consensus of the Committee was that input of protected materials constitutes reproduction within the meaning given to that term by international conventions and national legislation on copyright (Article 9(1) of the Berne Convention and Article IV bis of the Universal Copyright Convention). At this stage the act of reproduction takes place when protected materials are encoded on such media as punched or magnetic cards and tapes and fixed in the internal memory of the computer. Consequently, the copyright owner's consent should be required for this act, otherwise his right would be infringed. Another argument in favour of authorization at the input stage is the moral right of the author since there is a danger of such infringements as omission of his name, distortion or mutilation of words due to technical faults, etc. Moreover, a contract concluded at the stage of input would solve many problems connected with different forms of the output or retrieval of protected materials.

The Committee generally agreed that when the output takes place in the form of hard copy print-out, it should be considered as reproduction (of course, both at the stage of input and output, this does not concern the works exempted from copyright protection under national legislation). More complicated is the situation where the output is affected by projection of stored material in the form of visual images on a screen or cathod ray tube or similar process. May this act be considered as public presentation, performance, communication of works by wire, etc. or would it be covered by the right of reproduction? Can this display be compared to mere reading in the library? The international conventions on copyright do not seem to give a clear answer to this question. In view of divergent opinions expressed by the experts, the Committee stated that there was a need for further in-depth study of the questions surrounding output, with particular emphasis on generally recognized exceptions from copyright. It seems that it is mainly in this respect that the Committee recalled the fundamental concept of copyright - control of utilization of works - and that the right of reproduction and the right to put into circulation usually implies the right to control the destination of reproductions.

4. The right of the authors to control the use of their works was also of great concern to the Committee while discussing the question of administration of rights. A large majority of experts were of the opinion that authorization to use protected works for storage in and retrieval from computers should be based upon contractual agreements or other negotiated licences. It was pointed out that granting of such freely negotiated licences may be arranged either on an individual

basis or through such mechanisms as collective administration of rights and clearing houses. On the one hand, it would be premature to introduce compulsory or legal licences in that field at this stage, on the other hand, international conventions envisage such licences for particular categories of works and in special circumstances only but not in general.

However, some experts were strongly in favour of a system of compulsory or legal licences in this field. Otherwise, in their opinion the easy dissemination of works by computer systems, specially by minicomputers, cannot be achieved because of the difficulty for users of computers to obtain negotiated licences either on an individual or collective basis.

5. Finally, the Committee stated that within the context of a rapidly developing technology and increasing worldwide use of computers, it was highly desirable that recommendations for solution of copyright problems be made as soon as possible to provide some guidelines for national legislatures. In the meantime the Committee suggested that the international non-governmental organizations concerned take the matter of the use of computers for access to protected works under consideration and strengthen their efforts to safeguard authors' rights.

As asked by the Committee, the Secretariats of Unesco and WIPO are preparing, in consultation with the officers of the Committee, a draft Recommendation on the subject. The draft will be sent to all Member States for comments and later redrafted in the light of the observations received. The resulting draft will be submitted for discussion (and possible adoption) to the Second Committee of Governmental Experts to be held in 1982.

Notes

(1) Problems arising from the use of computers for creation of works which may enjoy protection under the copyright law and protection of computer software are not dealt with in this paper.

(2) See: "Forum", Council of Europe Quarterly, No. 1/81, p. XII

(3) It should be mentioned that, at the request of Unesco and WIPO (World Intellectual Property Organization) a number of extensive studies on the subject were carried out by Professor Eugen Ulmer of the Max Planck Institute in Munich.

G. EDITORIAL POLICIES AND NOMENCLATURE

EDITORIAL POLICIES

James LEQUEUX
Editor in chief, Astronomy and Astrophysics
Observatoire de Meudon 92190 Meudon FRANCE

ABSTRACT

A short summary of the way the main astronomy journals are produced is given. The different ways of publishing a scientific journal are described, their advantages and drawbacks are discussed and some projections towards the future are made. We finally deal with a few specific editorial topics : publication of large collections of data, designation of astronomical objects, indexing and retrieval of information.

This paper reflects my opinions concerning what a journal like Astronomy and Astrophysics is presently doing, could reasonably do and will possibly do in the future in order to publish the material submitted by the astronomical community. I have had contacts with Prof. B.J. Tayler, the Managing Editor of the Monthly Notices of the Royal Astronomical Society, and some of his opinions are conveyed here.

1. PRESENTATION AND COMPARISON OF THREE MAJOR JOURNALS OF ASTRONOMY AND ASTROPHYSICS.

There are a large number of journals in astronomy and astrophysics, some of them being highly specialized (e.g. Celestial Mechanics, Solar Physics) or containing only review papers (e.g. Annual Review of Astronomy and Astrophysics, Space Science Reviews). It would make no sense to compare journals which are not comparable. I thus decided to restrict the comparison to the three general journals which according to the Science Citation Index have presently the largest impact on the development of our Science : Astronomy and Astrophysics(Europe). The Astrophysical Journal (USA) and Monthly Notices of the Royal Astronomical Society (U.K.).

The following table gives the basic information concerning these journals. The information on the letter section is merged with that on the Main Journals ; however I found it useful to separate the information on the Supplement Series, which publish mainly data : this information is

C. Jaschek and W. Heintz (eds.), Automated Data Retrieval in Astronomy, 187–191.
Copyright © 1982 by D. Reidel Publishing Company.

given in italics when different from the corresponding one on the Main Journal. Monthly Notices has no more Supplements but publishes data on microfiches inserted in the Journal.

	ASTRONOMY AND ASTROPHYSICS + *SUPPL. SERIES*	MONTHLY NOTICES OF THE R.A.S.	ASTROPHYSICAL JOURNAL + *SUPPL.SERIES*
METHOD OF PRODUCTION	Composition, Camera-Ready from author *Camera-ready* from printer.	Composition *Microfiche*	Composition *Composition*
MICROFICHE EDITION	Yes *No*	No *(yes)*	Yes *Yes*
Nr OF PAGES IN 1980	4540+180(letters) *1600*	3800+275(letters) *4 microfiches*	8690+1385(letters) *1840*
PAGE CHARGES	No for Europeans yes for others	No	Yes
EDITORS IN CHIEF	2 + 1 (letters)	Decentralized Editorial Board	1 + 1 (letters)

I now summarize in the following table the advantages and drawbacks of the different possibilities for producing scientific journals.

WAY OF PRODUCTION	ADVANTAGES	DRAWBACKS
COMPOSITION	- Nice looking ; many possibilities(types, formulae)	- Expensive ; manuscripts are typed at least 2 times.
CAMERA-READY	- Inexpensive (≃ 2,5 times less as composition) ; manuscripts typed only once with modern text-processing facilities. Author responsible for text.	- Less nice-looking, limited possibilities compared to composition. Burde lies on authors if have only standard typewriters
MICROFICHES (always from camera-ready manuscript).	- Very inexpensive(≃15times less as composition). Takes little room on shelves.	- Need reader /copyer. Long-term behaviour may be bad. Authors generally very reluctant.

In view of the previous table, I think that classical composition from a typewritten text prepared by the author is likely to disappear in a not-too-remote future. However a possible way to save the advantages of composition would be that the printer makes the composition from a magnetic tape supplied by the author. This is in principle already possible but the lack of standardization of existing equipments hampers an extensive use of this system; camera-ready production may be a more satisfactory solution, although somewhat heavy for those authors who do not have access to a modern text-processing facility. Several astronomy institutes are however presently equiped with such facilities which allow easy corrections, margin justification, etc.. The softwares for producing formulae need improvements, but the situation is rapidly evolving. The price of a machine suitable for scientific editing is in the range $ 20 000 - 30 000. Camera-ready production by the author is meeting with some success at Astronomy and Astrophysics. The microfiche way is apparently accepted by astronomers only with extreme reluctance in spite of its enormous advantages, mainly for psychological reasons (an author likes to see his/her work in print !).

In the long run, it may well be that most or all of the scientific information will go on central magnetic-support libraries which can be interrogated by computer terminals. A commercial solution has already been developed in the USA. Although tempting, this system might not be accepted for the same psychological reasons as microfiches, and is particularly unfair for underveloped countries. Moreover halftone figures and even graphs correspond to large quantities of information which cannot be stored and transmitted as easily as the texts themselves; this serious technical problem has to be overcome before centrally computerized journals can be considered as acceptable.

I now discuss more specific points of interest in connection with the subject of the Colloquium.

II. PUBLICATION OF LARGE BODIES OF DATA

The publication of large catalogues, collections of pictures or drawings etc.. is a very expensive thing even if as usual the material is presented by the author in a form ready for publication. Already we tend to reduce the size so that the types are at the limit of readibility with the naked eye (microprint of 0.7 mm size); Astronomy and Astrophysiks also asks for a financial contribution from the authors if their paper is very long (say larger than 50 printed pages).

Should we stop publishing big catalogues ? Certainly not since they most often contain the basic data of our science. The question is rather : to which extent should catalogues be published in printed form ? As I said earlier, it is the experience of Prof. Tayler and myself that the authors are reluctant to publishing on microfiches. Data centers as the CDS in Strasbourg offer a better solution ; however they are not a universal panacea. Not every institute is connected by a terminal to the CDS, and sometimes interrogation of the computerized files is an heavy process if one only wishes to retrieve partial information on a few objects : in these cases, a catalogue on paper is certainly more handy. Finally information like graphs,maps, halftones, complicated formulae cannot be put on magnetic tape, and we find here the same problems as for computerized journals in general. It seems that for a long while a large fraction of the data will still have to be published on paper (or possibly microfiches) in spite of the heavy cost of the process.

III. DESIGNATION OF ASTRONOMICAL OBJECTS

Astronomical designation has always been a case for confusion as discussed elsewhere in this symposium. Practical solutions are presently elaborated in which any fixed astronomical object will have an unambiguous designation, probably based on celestial or galactic coordinates, which will supplement the present common designation(s). The Editors of Astronomy and Astrophysics and of Monthly Notices are certainly willing to comply with the new rules by making sure that any object cited in the papers has an unambiguous designation. They are also willing to recall as foot notes the meaning of abbreviations of the catalogues of celestial objects used in the papers they publish. As a concrete proof of this good will, may I recall that Astronomy and Astrophysics is financially participating (with IAU) to the publication of the Catalogue of the Nomenclature of Celestial Objects by Fernandez, Lortet and Spite, which will be a special issue of the Supplements.

However one should realize that checking that the authors actually comply with the new rules and/or helping them to do so will represent a rather formidable task for the editorial offices. We simply do not have at present the personel for doing that, and I can only be skeptical about the possibility of having the rules applied strictly in a near futur

IV. INDEXATION OF ASTRONOMY PAPERS, KEY WORDS.

It is clear that indexation of papers (generally made through the use of key words) is essential to the retrieval of the scientific information they contain. Unfortunately the situation in this respect is one of complete anarchy. For example, Astronomy and Astrophysics and The Astrophysical Journal each have their own thesaurus of key words ; both are different from the one used e.g. by Astronomy and Astrophysics Abstracts, etc.. Monthly Notices has no official one !

This situation can be explained by several considerations i) A thesaurus has to be adapted to some specific need and should be continuously evolving to follow the development of science ; for example, a list of key words built for preparing an annual subject index is not adapted to the needs of an Abstract journal which has to restrict severely the number of keywords per paper ; conversely, a scientist doing a bibliographic search in a specific field would like to retrieve small pieces of information hidden in many papers and prefers a very extensive list of key words for each paper, ii) No thesaurus is proven to be definitely better than the others. The sub-thesauri for astronomy included in the general physics ones are usually not detailed enough even for the needs of an Abstract journal.

I have not magic solution to offer to this problem, which however might not be as serious as one would think at first glance. I fear that no change, even if decided by a scientific body as representative and respected as IAU , would be universally accepted ; the lists of key words that each Journal uses have been elaborated through years of practice of the successive editors, and are used by the printer in a computer program for preparing the subject indexes ; any major change at the printer's office looks as a big affair when seen from the Editor side !

V. CONCLUSION

I am afraid that the present paper has raised many more questions than it has solved. I sincerely hope however that it will provide the astronomical community with a few basis for its thinking on the major item of scientific publication.

AIDS TO THE RETRIEVAL AND EVALUATION OF ASTRONOMICAL DATA

G. A. Wilkins
Royal Greenwich Observatory
Herstmonceux Castle, Sussex, England

ABSTRACT

Authors, editors and data centres should ensure that data on astronomical objects and systems can be readily located and that their accuracy can be properly judged by the user. In particular, authors and editors should endeavour to ensure that the title, abstract and other retrieval aids are accurate and informative and that the procedures used are described in sufficient detail to allow a proper evaluation of the results. Data centres should, where appropriate, provide additional aids to allow the users to find data efficiently and should, in all cases, ensure that each datafile contains adequate information about the sources of the data and about the significance of the fields in the records of the file.

INTRODUCTION

The main purpose of this paper is to draw attention to some of the ways by which authors, editors and data centres may significantly increase the probability that published data will be found and used by those who require such data. The paper also contains some information that may be helpful to those who wish to retrieve astronomical data. The adoption of the practices recommended here would lead to reductions in the amount of unnecessary observing and computing that now takes place because the astronomer has either failed to find already published data or has not been able to judge whether the data are of adequate quality for his purposes. Their adoption would lead to an increase in the amount of effective use of the data and a decrease in the amount of unnecessary publication and storage. More importantly, their adoption could lead to more rapid advances in our knowledge and understanding of the universe.

For simplicity of presentation we suppose that an astronomer wishes to find data about stars of a particular type, but similar considerations apply to searches for information about other types of astronomical object or system or technique. We are here concerned only with data

C. Jaschek and W. Heintz (eds.), Automated Data Retrieval in Astronomy, 193–198.
Copyright © 1982 by D. Reidel Publishing Company.

that are published in printed form and/or on magnetic tape. We recognise, however, that the published information may only refer to a source of further data about the stars, and that these data may be in non-numerical form, eg a photographic plate or as traces from a chart-recorder. The same principles apply: the information given should be sufficient to allow the astronomer to obtain access to the data and to judge whether they will be suitable and of sufficiently high quality to justify the effort involved in obtaining them.

AIDS TO RETRIEVAL

There are two phases in the retrieval of data, namely "search" and "fetch": firstly, it is necessary to establish where the data may be found and, secondly, it is necessary to obtain the data from that place, which may be in a journal, or on a magnetic tape, or in the store of a computer. In a fully automated system the fetch-phase will be a trivial operation once the search-phase has been successfully completed, but this is not yet the usual situation.

The most commonly used search technique is probably that in which the astronomer starts from his own knowledge of relevant material and then expands his search by looking up appropriate references given by each paper. This technique can involve much wasted time and expense unless the following conditions are satisfied: (a) each reference should include the title of the paper, a complete and unambiguous statement of the title of the journal, and the limiting page numbers, not just the number of the first page; and (b) the title of the paper should give a clear indication of its contents. The astronomer making the search can use clues given by the original author, but these may not be sufficient to avoid a wasted fetch when the reference is incomplete or the title is misleading. Review papers and the triennial reports on the progress of astronomy published in the Transactions of the IAU are very useful starting-points for such searches.

The second commonly-used search technique involves the use of abstracting journals and other published indexes, such as the Science Citation Index. Such searches may be carried out manually or by computer techniques since the contents of many abstracting journals are stored in computer databases. Astronomy and Astrophysics Abstracts provides the most useful single source of information about the current literature of astronomy, but it is not yet available in a database. Other English-language series containing abstracts of astronomical interest include Physics Abstracts and Scientific and Technical Aerospace Reports (STAR); these are accessible by computer techniques. Astronomical Abstracts are published in Russian in Referativnyi Zhurnal, part 51, and in French in Bulletin Signaletique, part 120.

The effectiveness of this technique depends on the quality of the information given in the entries as well as on the facilities provided in the abstracting journal or by the computer program. In particular the author should provide information in the abstract that will show the nature of the contents of the paper more clearly than the title is able to do; in particular he should indicate whether or not the paper contains observational data and results. In some disciplines, but not in astronomy, there are formal procedures for the "flagging" and "tagging" of data to indicate their presence and facilitate their retrieval (CODATA, 1976). The principal results or conclusions should be given in the abstract wherever this is practicable.

The arrangement of the abstracts in the journal and the supplementary indexing facilities that are provided also affect the effectiveness of searches. Unfortunately there are several different classification systems in use in astronomy. Since astronomers need to use information from other disciplines it appears to be extremely desirable that the Universal Decimal Classification should be adopted as the standard classification scheme for astronomy. Revised schedules for astronomy in class UDC 52 were prepared by an IAU/FID Joint Working Group quite recently (BSI, 1977). The UDC system usually allows the contents and format of a paper to be specified clearly and compactly and is quite suitable for use in information-retrieval systems. A guide to use of UDC in astronomy is in preparation at the Royal Greenwich Observatory.

At present many information-retrieval systems make use of keywords to describe the contents of a paper. A list of astronomical terms that are useful for this purpose has been prepared by the editors of Astronomy and Astrophysics Abstracts (Schmadel, 1979), and an extended list is being prepared by a Working Group of IAU Commission 5 (Lantos, 1981); this list should be consulted by editors and others responsible for the assignment of keywords. A multi-lingual version of this list would be useful. Some systems do not use assigned classification numbers or keywords but rely on scanning the title, and possibly the abstract, for words chosen by the person making the search. This should be borne in mind by authors when choosing titles for their papers.

Neither of the main techniques so far described is suited to the common astronomical requirement of finding data about, say, a set of stars satisfying specific selection criteria since, for example, all the relevant information cannot be given in the reference or abstracting journals. Instead, it is necessary to search the numerical datafiles themselves, and not merely bibliographic datafiles. Such searches are now usually carried out by computer techniques since many printed astronomical catalogues have been transcribed to cards or magnetic tape and most new catalogues are prepared by using computers. Datafiles should be made available in both printed and computer-readable form; extensive datafiles may be issued conveniently and economically on microfiche; magnetic tapes should be made available in standard formats.

The Stellar Data Centre at Strasbourg and its associated institutions have built up an extensive collection of astronomical datafiles and provide associated services. Information about these datafiles and services is distributed regularly in the Bulletin d'Information du Centre de Donees Stellaires (Observatoire de Strasbourg). Two of the datafiles deserve special mention because of their great value in searches for data: they are the Catalogue of Stellar Identifications, which provides cross-references between the names used for individual stars in different catalogues, and the Bibliographical Star Index, which aims to provide bibliographic references to papers giving information about individual stars. Even so the variety of astronomical objects and the bewildering number of different ways in which they are identified and by which their properties are described means that astronomers will continue to need patience, skill and a good knowledge of the literature of astronomy if their searches for published data are to be effective and efficient. Authors should follow the recommendations of the International Astronomical Union (1980) and ensure that the astronomical objects to which their data refer are unambiguously identified; they should seek the advice of the Union before introducing new styles of designation for new or unusual types of object.

A directory of sources of astronomical data has been compiled by Jaschek (1977) and republished by CODATA (1980) with amendments and additions.

Much useful astronomical data is not published in journals but in special series of reports issued by the observatories themselves. It is important that such reports be sent to the Astronomisches Rechen-Institut in Heidelberg so that they are included in Astronomy and Astrophysics Abstracts. A union list of the holdings of the principal astronomical libraries in the United Kingdom is in the early stages of preparation so as to facilitate the fetch phase of the retrieval of older material.

The original plates or other records invariably contain more information than the published reports of the observations. The retrieval of valuable information from such sources will be greatly facilitated by the compilation by IAU Commission 5 (Hauck, 1980) of a list of the plates, etc, that observatories hold and are prepared to make available for re-examination and re-measurement.

AIDS TO EVALUATION

The adoption of the following principles would make possible the proper evaluation of published data obtained from astronomical observations.

1. The paper must contain an adequate description of the observational procedures used to obtain the numerical data.

2. The paper must contain an adequate description of the procedure used to derive the reported results from the actual measurements.

3. The paper must contain full numerical results in a form that is as free from interpretation as possible and in such a manner that (a) the uncertainties of the data can be independently assessed and (b) the observational data can be re-analysed in terms of a hypothesis that is different from that considered by the author.

4. The data must be presented in such a way that the objects, system or phenomena observed and the quantities tabulated can be unambiguously identified and so that the results can be readily related to other data for the same or similar systems.

These principles have been developed from the guides prepared by CODATA (1973 & 1979) on the presentation of data derived from experiments and from observations in the geosciences. A full explanation of these principles, with specific examples of their application in astronomy, has been prepared (Wilkins, 1981) for submission to the IAU for approval prior to its publication by CODATA.

Similar principles apply to the publication of data derived by combining the results of different series of observations. The value of such data is very much diminished if the author does not provide adequate information about the sources of the original data and about the methods used to combine the data. Results based on computations for theoretical models must also be accompanied by sufficient information about the formulation of the model and the numerical values of the parameters to allow another person to reconstruct the computation and verify the published data.

Data centres should also ensure that the sources of the data that they store and distribute are specified with the data, preferably in a header block with information about the significance of the fields in the records of the file.

CONCLUSIONS

The advice given in this paper is not original but it is hoped that its presentation in this form will be found to be helpful to authors, editors, and the users of the astronomical literature. The costs of obtaining and analysing astronomical data are often very great and so it is clearly desirable that the published data should be presented in such a way that they can be readily found and fully utilised by those who are interested in them. The sourcebook prepared by Rossmassler & Watson (1980) contains several articles that are relevant to the retrieval and evaluation of data in the wider context of science and technology.

REFERENCES

BSI, 1977. Universal Decimal Classification: UDC 52 Astronomy Astrophysics Space Research Geodesy. ii + 43pp. BS 1000 [52]: 1977. British Standards Institution, London W1A 2BS.

CODATA, 1973. Guide for the presentation in the primary literature of numerical data derived from experiments. CODATA Bulletin No. 9, 6pp.

CODATA, 1976. Flagging and tagging data to indicate its presence and facilitate its retrieval. CODATA Bulletin No. 19, ii + 22pp.

CODATA, 1979. Guide for the presentation in the primary literature of numerical data derived from observations in the geosciences. CODATA Bulletin No. 32, i + 6pp.

CODATA, 1980. CODATA Directory of data sources for science and technology. Chapter 3. Astronomy. CODATA Bulletin No. 36, v + 56pp.

CODATA publications are available from CODATA Secretariat, 51 Boulevard de Montmorency, 75016 Paris, France.

Hauck, B., 1981. Preliminary results of an enquiry for better access to plate vaults. Bull. Inform. Centre Donnees Stellaires No. 20, 68-71.

IAU, 1980. Resolutions concerning the identification of astronomical objects in publications and the abbreviations of catalogues are given in Trans. IAU 17B, 42-43.

Jaschek, C., 1977. Survey of existing facilities. See Jaschek & Wilkins, 1977, 205-279.

Jaschek, C., and G. A. Wilkins (eds), 1977. Compilation, Critical Evaluation and Distribution of Stellar Data. D. Reidel Publishing Company, Dordrecht-Holland.

Lantos, P., 1981. Draft IAU vocabulary. 18pp. Observatoire de Paris, 92190 Meudon, France.

Rossmassler, S. A., and D. G. Watson (eds), 1980. Data handling for science and technology: an overview and sourcebook. xvi + 184pp. ISBN 0-444-86012-6. North-Holland Publishing Company, Amsterdam.

Schmadel, L. D., I. Heinrich and G. Zech, 1979. Astronomy and Astrophysics Abstracts: Vocabulary 06.79, vii + 152pp. Astronomisches Rechen Institut, Heidelberg.

Wilkins, G. A. 1981. Guide to the presentation of astronomical data. Preprint, 9pp, available from Royal Greenwich Observatory.

A LIST OF STANDARD STAR DESIGNATION

F. Ochsenbein
C.D.S. Strasbourg
F. Spite
Observatoire de Paris Meudon

ABSTRACT

Some rules about star designation have been edicted by Commission 5 of the IAU : they are briefly recalled and discussed. The essential rule, to be kept in mind by every author, is : when quoting a designation of a star, always quote the complete bibliographic reference of the paper defining this star designation. "Standard" designations can be an exception to this rule : it is proposed to publish in the CDS Bulletin a list of designations considered as "standard" ; the authors would be relieved of the obligation of quoting the references of these "standard" catalogues and could concentrate on the task of referencing the other catalogues.

INTRODUCTION

Several contributors to this Colloquium mentioned the problem and difficulties presently encountered with the designations of astronomical objects : these difficulties have been known for years. A discussion took place at the Grenoble meeting of IAU, in 1976. Working groups circulated proposals before the Montreal meeting of IAU (Jaschek and Pecker, 1979, Fernandez et al. 1979). From the discussion in Montreal, a few recommandations were extracted and adopted : let us concentrate here on stellar problem.

1. VERY BRIGHT STARS

The recommandation (Jaschek et al. 1980) precise that Beyer and Flamsteed designations are to be prefered for the brighter stars. Since we are involved in bibliographical work, we remark that, from an empirical point of view, some very bright stars are systematically quoted in the literature by their latin (or arabic) names such as : Vega,

C. Jaschek and W. Heintz (eds.), Automated Data Retrieval in Astronomy, 199–202.
Copyright © 1982 by D. Reidel Publishing Company.

Sirius, Procyon, Arcturus, Pollux, Algol, Mira Ceti ... Since these designations are not ambiguous, we are ready to admit that these designations are acceptable if coupled with a second designations (number in a standard catalogue and/or coordinates). Here we encounter a principle, outlined by Prof. Jaschek (Jaschek et al., 1980) that a double designation is always extremely useful.

Commission 5 proposes Bayer designations. In our opinion, this should be considered with some caution : every astronomer who has ever done some proofreading of a paper including υ and ν, χ and κ, ξ and ζ immediately knows what we mean. Moreover, there are some ambiguities with Bayer designations in the Southern Hemisphere (Hoffleit, 1979) Admittedly, most of the problems come from superscripts, and it could be decided to retain only the Bayer designations which are without superscripts. But Bayer designations used by differents authors do not agree about the presence of superscripts. Finally the solution should be : use only the Bayer notation exactly as they are defined in the next edition of the Yale Catalogue of Bright Stars = BS = HR. Note that you have to look carefully about this star in this catalogue for checking carefully the Bayer designation : then you have at hand the Bright Star number : please quote this number too as a second identifier, thus solving any ambiguity problem.

Dr Hoffleit (1979) recalls that it was recognised, a long time ago, that the use of latin lower case letters was to be avoided. Only a few variable stars have no other clearly defined name : it could be a good time now to progressively overcome this inconsistency.

Commission 5 then recommands the use of Flamsteed numbers ; this designation has no real ambiguity problems. Let us note that one of the reasons for maintaining alive these older designations, is that they keep a link with observations in the past. May we remark that the limits on Constellations were changed by IAU, and that a few stars have officially lost their Flamsteed designation : in this case, the only way to maintain a link with past observations is to use more modern designations such as HR = BS or HD.

2. BRIGHT STARS

For stars a little fainter, Commission 5 recommands HR numbers : unfortunately, most people use BS instead for HR, and we must admit that for a young astronomer working in space physics, the logical link between a Bright Star number and the old Revision of Harvard Photometry is probably not evident.

3. INTERMEDIATE STARS

For intermediate stars, HD, BD, CD, CPD and Giclas numbers are convenient. However, rather than G., GD, and GR, the replacement of G by the full name Giclas is recommanded, in order to avoid confusion with other (stellar and non-stellar) astronomical designations. Again, for all these stars, publication of coordinates as a second identifier would be useful, if cost (or policy) of publication allows it.

4. FAINTER STARS

Here, the situation is difficult : it is necessary to rely on special lists. It is recommanded to publish the coordinates, where available Moreover, it is necessary to quote the complete bibliographic references of these special list.

5. ASTRONOMERS NAMES

A few stars are quoted by the name of an astromer.

If the star has a "standard" designation, then always quote also the standard designation. If the star has no "standard" designation, always quote the complete bibliographic reference of the paper identifying the star, and the coordinates.

We would not have insisted on this very small sample of stars, if errors were not often repeated in the modern literature. For instance, a book published in 1976 repeats the false statement that Barnard's star would be identical to BD 4°3561, in spite of continuous efforts of Prof. Gliese to point out this error.

A list of stars designated by astronomers'names will be published in a forthcoming CDS Information Bulletin.

6. VARIABLE STARS

The nomenclature of these stars is kept up to date by our colleagus in Moscow. We mentioned hereabove that probably lower case latin letters should progressively be replaced (in order to avoid errors due to misprints).

7. STARS IN CLUSTERS, IN MAGELLANIC CLOUDS

Nothing was proposed. For stars in clusters, Mermilliod (1977) proposed a special numeric code ; the IAU Commission of clusters edicted a nomenclature rule for clusters, materialized by the Catalogue of Lyngå (1980) : a solution could probably be found in the near future.

UNIVERSITY COLLEGE LIBRARY CARDIFF

8. CONCLUSION

The solution to be adopted, in order to completely clarify the nomenclature problems, are not easy to find. However three simple rules, if followed by all authors, could considerably alleviate the problems :

1.1. The simultaneous use of two designations.

1.2. The systematic citation of the complete bibliographic references of the designations (only the designation included in an accepted list of "standard" designations could be excepted).

1.3. The systematic indication of the equinox of the coordinates.

Let us hope that all the participants in this Colloquium will at least remember these three rules, follow them and teach them to their students and younger colleagues ; this would be a first step in bringing astronomical nomenclature to an orderly state, not spoiling any more the prestige of Astronomy.

9. FINAL MESSAGE

Compilations of stars, built up by specialists, are very important and very useful work. This is so important for bibliographical work and nomenclature problems, that anybody aware of a compilation in progress somewhere should inform immediately the authors of this communication.

REFERENCES

Fernandez A., Lortet M.C., Spite F. : 1979, "Compilation of Data for preparing resolutions about nomenclature rules", unpublished.
Hoffleit D. : 1979, Bull. Inform. CDS, 17, pp. 38-65
Jaschek C., Ochsenbein F. and Bidelman W.P. : 1980, Bull. Inform. CDS 18, pp 41-50
Jaschek C., Pecker J.C. : 1979, Bull. Inform. CDS 16, pp. 57-70
Lyngå G, Lundstrom I. : 1980, in IAU Symp. 85, 123 (D. Reidel Publ. Co., Dordrecht, Holland)
Mermilliod J.C. : 1977, "Principles of a Coded numbering System and its application to open clusters, in Compilation Critical Evaluation and Distribution of Stellar Data", IAU coll. 35, C. Jaschek and G.A. Wilkins, Eds, D. Reidel Publ. Co., Dordrecht, Holland.

DICTIONNAIRE DE LA NOMENCLATURE DES OBJETS CELESTES.

A. Fernandez, M.C. Lortet, F. Spite
Observatoire de Meudon, France.

Abstract. The confusing situation of the nomenclature of the astronomical objects led us to build, as a first step, a dictionary of the designations presently used. The future publication of this work is supported by Commission 5 of IAU. A brief description of this dictionary, as well as a sample of it, is given here. The designations (acronyms) will be given, together with the format, the total number of objects, the type of objects, the code for the complete bibliographical reference, the signification of acronym... Nomenclature for X and γ-ray sources, designations by constellations and descriptive designations will be dealt with as well. Good compilations are a considerable help in nomenclature problems ; every effort should be made in order of : 1) urging specialists to build compilations 2) announce compilations in progress to Astronomical Data Centers and to the present authors.

La situation présente de la nomenclature des objets célestes est loin d'être satisfaisante (Jaschek et al., 1980). Ce problème a déjà fait l'objet d'une discussion organisée à la dernière Assemblée Générale de l'IAU à Montréal par MM. Jaschek et Pecker. Des délégués ont été nommés pour chaque commission. Des documents ont été rassemblés et diffusés : compilation des résolutions prises dans le passé par l'IAU (Jaschek et Pecker, 1979), guide des désignations les plus usitées.

Un travail préparatoire de compilation des nomenclatures a été fait (Fernandez et al. 1979, publié par le CDS) basé notamment sur Lortet (1978), Spite et al. (1978), Collins (1977). Ce travail donnait, avec des références incomplètes, la liste d'environ 900 nomenclatures. A l'aide des remarques reçues de divers représentants de commissions, il est actuellement complété et mis en forme et deviendra le dictionnaire de la nomenclature des objets célestes.

DESCRIPTION DU DICTIONNAIRE.

Ce dictionnaire donne, sauf pour le système solaire, les désignations "réellement" rencontrées dans la littérature, c'est-à-dire qu'il exclut les noms de catalogues (ex. BGC : De Vaucouleurs et al. 1964 qui

C. Jaschek and W. Heintz (eds.), Automated Data Retrieval in Astronomy, 203–205.
Copyright © 1982 by D. Reidel Publishing Company.

cite les galaxies par leur nom classique) et les désignations "fictives" (ex. AMWW : Altenhoff et al. 1960) qu'on trouvait dans Collins 1977. Certaines listes d'étoiles dans les amas ont également été exclues. Il est composé de plusieurs parties :

1) Le dictionnaire proprement dit disposé en colonnes :

DESIGNATION	FORMAT	NOMBRE	TYPE D'OBJET	R	CODE	REFERENCE	SIGNIFICATION DE L'ACRONYME
A = Aitken			Double *			Use IDS Nomencl.	
A			Flare *				Asiago
A	N	8	SMC V. *		1965 A	Andrew. Use Variable * Nomencl.	
A			Double *			See Ads	
A	R	9	HI CLD		1973 G	Giovanelli +	
A = A'	NN	86	PN		1966 A	Abell. Use PK ref. 1967 P	Abell
A = Abel = Abell	NN	73	PN		1955 A	Abell. Use PK ref. 1967 P	Abell
A = Abel = Abell	NN	13	GCL		1955 A	Abell	Abell
A = Abel = Abell	NNNN	2712	Cluster of G		1958 A	Abell	Abell
A1	A1-NNNA		Faint blue				Asiago. field 1
			Objects				
AAH			Cluster *		1961 H	Hoag +	A.A. Hoag
AB		175	UV		1970 B	Braccesi, A. +	Bologna ?
Abel = Abell						See A = Abel = Abell (3 different lines)	
ABGC	NNNN	2712	Cluster of G		1958 A	Abell (See A = Abell)	Abell Galaxies Clusters ?
ABT		30244	*		1972 A	Abt +	Abt
ABT = Abetti	NN	23	Double *			Use IDS Nomencl.	
AC					1974 E	Described by Eichhorn	Astrographic Catalogue
AC	NN	20	Double *			Use IDS Nomencl.	Clark, Alvan
A.D. = AD	NNNN		SN			Use Variable * Nomencl.	Anno Domini
ADS	NNNNN A,B	17180	Double *			Ref and more modern data in IDS	
AE	N	1	V *			Use Variable / Nomencl.	A. Elvius, 1975
			He-rich				

colonne 1 : Acronymes en ordre alphabétique. Par acronyme, nous entendons la partie littérale d'une désignation (ex. PKS dans PKS 2302+28).
colonne 2 : Format, c'est-à-dire disposition et signification des chiffres (ex. NNN signifie un numéro d'ordre à 3 chiffres, HHMM ± DD fait intervenir les coordonnées équatoriales).
colonne 3 : Nombre d'objets du catalogue (ne signifie pas forcément numéro du dernier objet car il y a souvent des numéros manquants).
colonne 4 : Type d'objet ou longueur d'onde sous forme abrégée.
colonne 5 : Signale les remarques concernant le format.
colonne 6 : Numéro de code renvoyant à la référence complète ; il comporte l'année, l'initiale du premier auteur et un numéro d'ordre.
colonne 7 : Nom du premier auteur et éventuellement renvoi à une compilation ou à une autre désignation.
colonne 8 : Signification de l'acronyme.

2) La liste des références. Précédées de leur numéro de code, elles sont classées par année et, à l'intérieur des années, par ordre alphabétique. Le ou les acronymes se rapportant à chaque référence sont également indiqués.

3) La table des désignations et des références X et γ, mises à part à cause de leur renouvellement trop rapide.

4) Les désignations comportant un nom de constellation et leurs

références.

5) Les désignations descriptives (ex. Crab Nebula, Horsehead,...) et leurs références.

6) Des tables auxiliaires expliquant le format et les abréviations employées pour les types d'objets ou les longueurs d'onde.

COMPILATIONS.

Dans certains cas, des renvois à des compilations ont été pour nous un gain en temps et en place car elles donnent les références de nombreuses listes parfois anciennes que nous avons alors réduites au nom du premier auteur + l'année (ex. A. Acker pour les nébuleuses planétaires, G. Lyngå pour les amas ouverts, Kukarkin pour les étoiles variables,...). Ces compilations - dans le sens de listes systématiques, et aussi complètes que possible, d'un certain type d'objet et avec "cross-identification" (correspondance de la désignation) - nous seraient extrêmement utiles, sous forme publiée lorsque c'est possible. Elles pourraient par exemple être centralisées au CDS. Il serait extrêmement important que les compilations en cours nous soient signalées le plus vite possible.

CONCLUSION.

Il y a bien des difficultés dans la nomenclature astronomique ; soulignons deux ou trois règles qui pourraient améliorer la situation : Si un astronome publie une liste d'objets sans donner de désignation à sa liste et que cette liste se trouve être fondamentale, les objets qui la composent seront cités par d'autres astronomes. Il faut alors la citer en forgeant une désignation comprenant plus de deux lettres (il faudrait vérifier qu'elle ne coïncide pas avec une désignation déjà usitée). Le nom de l'auteur, les initiales des auteurs, le nom de l'Observatoire ou une indication du type d'objet sont acceptables s'il n'y a pas ambiguïté.

REFERENCES.

Collins, M.: 1977, Astronomical Catalogues 1951-1975, INSPEC Bibl. Ser. N° 2, Hitchin, Great Britain.

Fernandez, A., Lortet, M.C., Spite, F.: 1979, Compilations of data for preparing resolutions about nomenclature rules, Ed. CDS, tirage limité.

Jaschek, C., Pecker, J.C.: 1979, A Meeting on the Nomenclature of Astronomical Objects, Bull. Inf. CDS 16, pp. 57-70.

Jaschek, C., Ochsenbein, F., Bidelman, W.P.:1980, Joint Meeting on the "Nomenclature of stellar galactic and extragalactic objects", Montréal, Bull. Inf. CDS 18, pp. 41-50.

Lortet, M.C.: 1978, Propositions pour un centre de données galactiques non stellaires, Bull. Inf. CDS 14, pp. 79-104.

Spite, F., Kirchner, S., Lahmek, R.: 1978, List of stellar Catalogues used in the Bibliographic Star Index (BSI), Bull. Inf. CDS 15, pp. 2-30.

ON THE NOMENCLATURE OF DIFFUSE FRAGMENTARY OBJECTS

P. Polechova
Observatory Petrin, Prague, Czechoslovakia

ABSTRACT

For the purpose of cataloguing diffuse objects in a wide range of dimensions, two alternative proposals are presented.

Problems similar to those discussed in connection with the nomenclature of optical HII-regions may occur with any diffuse objects such as molecular clouds, large reflection nebulae, HI-regions, and, of course, radio HII-regions.

The radio HII-regions are most commonly designated by the galactic numbers "G". The galactic longitude forms the first part of this number, the galactic latitude the second part, both parts being usually accurate to the first decimal digit and separated by the sign of latitude. This system (also used sometimes for optical objects) was introduced by Mezger and Henderson (1967) in a survey at 5 GHz, with a half power beam width of the telescope of about 6'. The angular dimensions of their objects were in the range from 3 to 11'.

The angular diameters of the HII-regions stretch over a wide extent of at least three orders of magnitude - from several arc sec up to several degrees or tens of degrees. For larger objects, galactic numbers accurate to two decimal digits would be inconvenient since large absolute errors may occur in the coordinates, which would change the name. Galactic numbers accurate to one decimal digit are not usable throughout the catalogue, either since among the optical HII-regions (represented on the northern sky by the catalogue of Sharpless)there exists at least one triple (S 192, S 193, S 194) and one couple (S 256, S 257) of objects the galactic coordinates of which are identical for all the three or the two objects, respectively, if they are denoted with accuracy of the first decimal digit. On the other side there is an example of the radio counterpart of NGC 2024 for which Wilson et al. (1970) and Reifenstein et al. (1970) give G 206.5-16.4, while Schraml and Mezger (1969) give G 206.6-16.4. This example shows that the names should not be introduced freely by the authors.

C. Jaschek and W. Heintz (eds.), Automated Data Retrieval in Astronomy, 207–210.
Copyright © 1982 by D. Reidel Publishing Company.

I think the new objects should be named by running numbers with the mark of the discoverer firstly. The incorporation to the system of the general catalogue should then be made cautiously. The system should consider the following conditions.

It should be open to later discoveries, including both the discoveries of "new" emission centers and the discoveries of new extensions of "old" objects.

It should include rough information about the position in the Galaxy and the advantage of the galactic coordinates lies also in their constancy with time.

The number should not be regarded as a substitute of any kind of coordinates, and the once established name should not be changed.

Since generally a higher resolution of a sky survey implies a more detailed fragmentation of the objects, every survey of the sky can only be appropriate for a limited range of angular dimensions.

The coordinates in a general system may depend not only on the resolution but also on the frequency.

I think that the best workable system is one which uses and refines the system already in use. The result of such considerations is a proposal of a multiscale nomenclature system, where the information about the dimensions would be included in the name of the object, in addition to the information about its galactic coordinates. (Moreover, it can contain also information about the frequency in place of the, formerly constant, G.).

Each object can be affixed to a definite size of a coordinate box according to its angular diameter. Thus the appropriate accuracy of the coordinates (which determines the names of the object) would be assured. The correspondence of the objects' dimensions to the boxes is shown in Table 1. Table 2 presents an alternative which is a logical extension of Table 1.

In the range of the one-decimal coordinate boxes (former G-numbers) the catalogue of Rodgers et al. (1960) and for instance that of Bok et al. (1955) would be usable as a guide for the purpose of the general catalogue of the optical HII-regions.

Objects	small	medium	large
Size Θ of the objet	less than 6'	$6' \leq \Theta < 60'$	larger than 60'
Size of the box	0.01°	0.1°	1°
Name	LLL.LL±BB.BB	LLL.L±BB.B	LLL±BB

Table 1. The proposed rules for the range of the accuracy of the coordinates determining the names of the objects - alternative 1

Size Θ of the object	...	$3.6'' \leq \Theta < 36''$	$36'' \leq \Theta < 6'$	$6' \leq \Theta < 60'$	$60' \leq \Theta$
Size of the box	...	0.001°	0.01°	0.1°	1°
Name	...	LLL.LLL±BB.BBB	LLL.LL±BB.BB	LLL.L±BB.B	LLL±BB

Table 2. The proposed rules for the range of the accuracy of the coordinates determining the names of the objects - alternative 2.

REFERENCES

Bok B.J., Bester M.J., Wade C.M. : Daedalus 86, 9 or Harvard reprint N° 416 (1955)

Gum C.S. : Memoirs Roy. Astron. Soc. 67, 155 (1955)

Lynga G., Hansson N. : Astron. Astrophys. Suppl. 6, 327 (1972)

Marsalkova P. (=Polechova P.) : Astrophys. Space Sci. 27, 3 (1974)

Mezger P.G., Henderson A.P. : Astrophys. J. 147, 471 (1967)

Münch L. : Bol. Obs. Tonantz. Tac. 2 N° 13, 28 (1955)

Perek L., Kohoutek L. : Catalogue of Galactic Planetary Nebulae, Praha (1967)

Polechova P. : Ph. D. Thesis, Charles University, Praha (1979)
Rodgers A.W., Campbell C.T., Whiteoak J.B. : Monthly Notices Roy. Astron. Soc. 121, 103 (1960)
Reifenstein E.C. III, Wilson T.L., Burke B.F., Mezger P.G., Altenhoff W.J. : Astron. Astrophys. 4, 357 (1970)
Schraml J., Mezger P.G. : Astrophys. J. 156, 269 (1969)
Sharpless S. : Astrophys. J. Suppl. 4, 257 (1959)
Sivan J.P. : Thèse de Doctorat d'Etat, Université de Provence, France (1977)
Van den Bergh S. : Astron. J. 71, 990 (1966)
Walmsley C.M. : unpublished (1973)
Wilson T.L., Mezger P.G., Gadner F.F., Milne D.K. : Astron. Astrophys. 6, 364 (1970)
Wilson T.L., Downes D. (Ed.) : Lecture Notes in Physics 42, HII-regions and related topics, Springer-Verlag (1975)

THE CATALOGUE OF STELLAR IDENTIFICATIONS

Ochsenbein F. - Bischoff M.
Centre de Données Stellaires
F-67000 Strasbourg

SUMMARY

The evolution of the Catalogue of Stellar Identifications (C.S.I.) developed by the Stellar Data Center (C.D.S.) is summarized in Table 1 ; a microfiche edition of this fundamental file is described in another paper (Ochsenbein et al, 1981), and recent developements of this file are presented.

The Catalogue of Stellar Identifications (C.S.I.) has now a rather long story, since it was started ten year ago by J. Jung and M. Bischoff at the very beginning of the Stellar Data Center (C.D.S.). As indicated by its name, the C.S.I. essentially deals with the stellar nomenclature, in the sense that the first aim of this catalogue is to provide the different names or identifications for each star. Once such a dictionnary of stellar designations is built up, and stored in a computer, the question of finding the whole literature about a star becomes realistic (see the presentation of the B.S.I., Bibliographical Star Index, at this Colloquium) ; it becomes also much easier to link several catalogs in order to get more information about each star.

The fundamental role played by stellar designations and the difficulties related to the inextricable nomenclature system in use (see e.g. Jaschek et al, 1980) lead us to develope the C.S.I. A more complete presentation of the C.S.I. and of its possibilities may be found elsewhere (e.g. Ochsenbein et al, 1981) ; let us just say that an in-line access to the C.S.I. now allows to get immediately the informations about positions, proper motions, parallaxes, photoelectric photometry in 6 systems, MK classifications, radial and rotational velocities, H γ equivalent widths and bibliographic references for a star.

Table 1 summarizes the evolution of the C.S.I.

C. Jaschek and W. Heintz (eds.), Automated Data Retrieval in Astronomy, 211–213.
Copyright © 1982 by D. Reidel Publishing Company.

Year	Stars	Catalogs	Identifications
1972	380 000	9	1.11 10^6
1976	440 000	28	1.38 10^6
1979	440 000	36	1.75 10^6
1982 ?	(470 000)	(200)	(2 10^6)

The first version of the C.S.I., in 1972, was a merge of some fundamental catalogues that did exist on machine readable form. We tried essentially, in a second step, to enter into the C.S.I. the catalogues which are essential in astrophysical studies : UBV catalogues, MK classification lists, etc... ; this step was still in progress in 1976, and we feeled in 1979 that, with 36 connected catalogues including the most important data in the classifical fields of stellar astronomy and a mean number of 4 entries per star, an edition on microfiche of this catalogue with the known data would be useful to ensure a larger diffusion (Ochsenbein et al, 1981).

A third step was started about one year ago, and began with a comparison between C.S.I. and B.S.I. : in recent years, more and more stars were entered into the B.S.I. with new designations. About 200 catalogues are recorded in the B.S.I., while only 36 are known in the C.S.I. This difference means that newly discovered interesting stellar objects are missing from the C.S.I., and we feel that this gap must be filled.

But incoporating 150 more catalogs in the C.S.I. is a painful and time-consuming task, since all these lists must be checked by hand. One of the authors (M. Bischoff) has undertaken this extension of the file. At the present time, 46 lists including 21 256 stars have been checked : among there 21 256 stars, 8861 were already included in the C.S.I., which means in practice that 42 % of the stars with exotic designations are previously identified stars.

We also plan to add new cross-identifications between CoD and CPD which have been built up in the zone -22° to -40° by Bonnet (1981) ; as a final result, we hope that at the end of 1982 (Table 1) the number of stars will reach 470 000 with 200 connected catalogues.

Finally, it is a pleasure to thank all people who contributed to this fundamental file - that is not only the staff members of the C.D.S., but also astronomers who contributed by sending valuable catalogues or errata lists. It is also a pleasure to learn that the C.S.I. is used in many places - even if we feel that our efforts are becoming more and more an anonymous contribution to studies in stellar astronomy.

REFERENCES

Bonnet R., Ochsenbein F. : 1981, C.D.S. Inf. Bull. 21, 51

Jaschek C., Ochsenbein F., Bidelman W.P. : 1980, C.D.S. Inf. Bull. 18, 41

Ochsenbein F., Bischoff M., Egret D. : 1981, Astron. Astrophys. Sup. 43, 259

H. DATA IN ASTRONOMY

DATA IN ASTRONOMY

B. Hauck
Institut d'Astronomie de l'Université de Lausanne
CH-1290 Chavannes-des-Bois

Abstract.- The present situation regarding data for astronomy and astrophysics is dominated by two factors: large increase of data and better accessibility, mainly due to the existence of data centres. An attempt to centralize the information of plate vaults is also presented.

1. INTRODUCTION

It is perhaps unnecessary to mention before an astronomers' assembly the importance of data in astronomy and astrophysics, especially if these astronomers are concerned with data. We are certainly not the first to be occupied with astronomical data, since the first star catalogue was established by Hipparcos. Nevertheless, in Hipparcos' time, during the second century B.C., there was no need to speak about data centres: 1080 stars formed this first catalogue. The development of star catalogues was initially due to growth of astrometric data, and once physical methods were used, catalogues with physical data were also compiled. A good example is undoubtedly the Henry Draper Catalogue.

Today also data play an important role in astronomical and astrophysical studies. Catalogues are not always the only result of observations but also a compilation of data for the preparation of an important experiment. To take an example, in a few years ESA will launch an astrometric satellite, Hipparcos, the purpose of which is to obtain data for more than 100,000 stars. At the present time ESA is asking who will be willing to prepare the "Input Catalogue", i.e. the catalogue of all the main data for the programme stars!

Fortunately for people having need of astronomical data, many astronomers are interested in the compilation, storage, etc. of these data. An inquiry made by C. Jaschek (1980) and published by CODATA as the third chapter of the CODATA directory of data sources for science

C. Jaschek and W. Heintz (eds.), Automated Data Retrieval in Astronomy, 217–225.
Copyright © 1982 by D. Reidel Publishing Company.

and technology reveals that 70 centres or groups of various dimensions are in existence. All the astronomical activities are covered, but not all to the same degree. Many centres - in fact 35 - are concerned with the Sun and the solar system, 35 also with stars and star systems, 4 with various items and only one with radio observations and also only one with galaxies' data. I am not sure that data and data compilation play the same role in each field, but I have the feeling that there is certainly a need for a data centre for non-stellar objects. However, the management of a data centre can only be done by specialists in that particular field and I hope that the initiative will come from colleagues concerned with non-stellar objects.

Data are also important in other sciences and there are in fact many problems in common. Thus, in order to improve the quality, reliability and accessibility of data of importance to science and technology, the International Council of Scientific Unions (ICSU) established in 1966 CODATA, the Committee on Data for Science and Technology. The IAU is represented on CODATA, but mainly with regard to administrative matters, and it is important that all astronomers concerned know of the existence and purpose of CODATA with a view to participating more intensively in its scientific activity. The scientific endeavours of CODATA are largely achieved by task groups, of which I can mention those perhaps related to our science: Accessibility and Dissemination of Data, Computer Uses, Fundamental Constants, Space and Time-dependent Data (of which the chairman is an IAU member), International Training Courses on the Handling of Experimental Data. Under the sponsorship of CODATA and UNESCO a book was published recently entitled Data Handling for Science and Technology, by S.A. Rossmassler and D.G. Watson (1980) and many astronomers would benefit by reading it. My hope is that in the near future more astronomers will collaborate in the activity of CODATA.

2. GROWTH OF DATA

When we consider the various problems related to astrophysical and astronomical data we see that a main problem is the growth of these data. This reflects the development of the traditional activities in astronomy and astrophysics. Woltjer (1978) has shown how the number of large telescopes has grown during this century. During the first half, only one telescope with an aperture of more than 2.5m was in existence until the completion of the Hale telescope in 1948. Now 15 big telescopes are in operation or under construction. The majority of these telescopes have only recently started to produce a large amount of data and it is thought that the growth of data will be even more important in the near future. This growth is also strengthened by data obtained by satellites, which have already produced a large amount of data. The main problem here is perhaps the large quantity in a short period of time. For example, the IUE satellite has obtained 16,000 spectra in two years and two months. The data collected by ESA and SRC (one-third of the total) is stored on 500 tapes of 1600 bpi!

The growth of data in stellar astronomy was studied recently by C. Jaschek (1978). In some cases (trigonometric parallaxes, visual binaries) this growth is decreasing, whilst in others it is increasing, sometimes exponentially (variable stars, asteroids, orbits, *uvby*β and Geneva photometries). New catalogues (Hauck and Mermilliod 1980, Rufener 1981) for both the above-mentioned photometric systems have been published since Jaschek's paper and the amount of data in these catalogues confirms his results. For the Geneva system the number of stars passes from ~5000 to ~15,000 between 1976 and 1980! (see table 1). However, the situation is perhaps not so optimistic. 15,000 stars represents only 6.6% of the HD catalogue and if we are interested in a special sample, many photometric data are likely to be absent. But the situation is not always bad, for example less than 10% of the stars belonging to the Gliese Catalogue of Nearby Stars have no photometric data at all (Hauck and Mermilliod 1981).

Table 1. Growth of photometric data

A. *uvby*β system

1965	1.2	10^3
1973	7.5	10^3
1975	9.3	10^3
1977	1.44	10^4
1978	1.9	10^4

$N=1238e^{0.21\Delta t}$ $\Delta t=t-1965$

B. Geneva system

1964	3.4	10^2
1966	6.8	10^2
1971	1.4	10^3
1976	4.6	10^3
1978	1.3	10^4
1980	1.5	10^4

$N=345e^{0.24\Delta t}$ $\Delta t=t-1964$

3. CRITICAL EVALUATION OF DATA

Critical evaluation of astronomical and astrophysical data is certainly a tedious job, but very important for all our community. In stellar astronomy the progress situation is very different from one field to another. With regard to variable stars and parallaxes we need a new catalogue in both cases and hopefully these catalogues are in preparation respectively at Moscow and Yale Observatory. For radial velocities the situation is definitely worse, but it is good again if we consider star positions, spectral types and photometric data.

The computer can play an important role in the compilation and homogenization of data. We can use it to obtain weights by intercomparison of lists and thus we are able to calculate a weighted mean as homogeneous data. But we can also keep only one set of data per star by giving the order of preference amongst all the references. Such a procedure was recently applied by M. Jaschek (1978) to stars having an MK type. The resulting catalogue contains ~30,000 stars. This work is based on the La Plata Catalogue (C. Jaschek et al. 1964) and the regularly updated compilation of Mrs. P. Kennedy. Another effort in this direction may also be mentioned, that made by W. Buscombe (see this colloquium). In photoelectric photometry we apply the other procedure as described by Nicolet and Hauck (1977). The most important work published during these last years is without doubt Nicolet's catalogue of UBV measurements. This catalogue has the merit of giving UBV homogeneous data for 53,845 stars. In addition, an indication of quality is given for the V magnitude, U-B and B-V colour indices of each star. This work was possible only by using a computer and because the previous compilations of Blanco et al. (1964) and Mermilliod and Nicolet (1977) were also available on magnetic tape.

It is certainly difficult to recommend such and such a method to homogenize data. The one to be used will depend on the field. Nevertheless, in all cases it is necessary for the reader to have access not only to homogeneous data but also to primary data or, in other words, it is necessary to be able to identify the sources of the compiler. This is also very easy by using a computer.

4. ACCESSIBILITY OF DATA

The accessibility of data is certainly a major problem today. Literature and scientific journals play an important role but cannot assume the full burden. Data centres are definitely a good solution but nevertheless only a partial one. The merits of the existing centres are numerous and I wish to mention especially the fact that their existence is possible since many data files are in machine-readable form. This existence has in turn encouraged many authors to put their data on tape. The fact of having data in machine-readable form is a good step to better access to the data. Another merit of data centres is the preservation of data, their better distribution and the possibility of treating various problems more easily, for example those linked to the various identifications for the same star.

The present data centres in our field are relatively well located and certainly of sufficient number. Nevertheless, the situation is not perfect. Firstly I think that it should be improved by stronger collaboration between astronomers obtaining data and those compiling these data (and the improvement should be in the sense that the data should be sent to the compilers in machine-readable form). Secondly, we are hoping that data networks will be more numerous in the near future. Starlink and the Japanese network are certainly good examples, but with a limited (geographically speaking) interest. It is my hope that astronomical data networks develop on a wider basis.

In figure 1 I present an organigram of links between the acquisition, homogenization and distribution of astronomical data. At the present time it is only a prospective view, but I am nevertheless confident that we can realize this plan without too much difficulty but with a large amount of goodwill. Thus we will have contributed to improving accessibility to astronomical data.

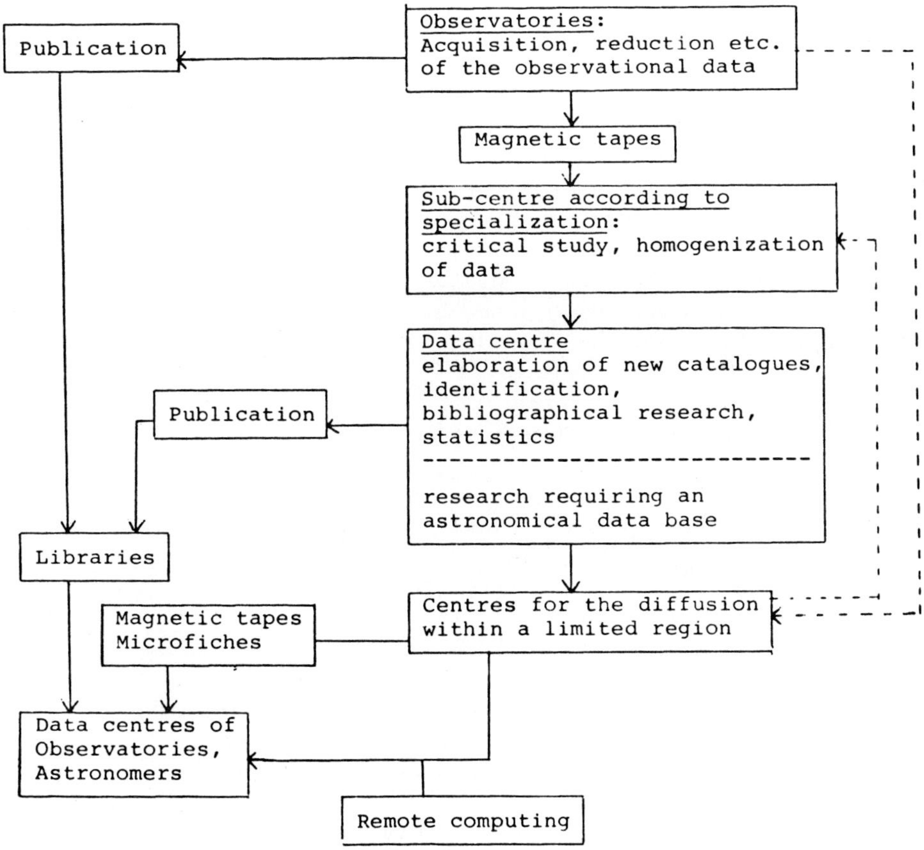

Figure 1. Organigram of links between the acquisition, homogenization and distribution of astronomical data

Our care to improve accessibility to the data has led to the discovery that there was an important problem concerning the plate vaults. Many plates are stored in various observatories and it is difficult to know which spectra or which fields are available. The Working Group on Astronomical Data of IAU Commission 5 has examined the possibility of improving access to plate vaults. From our point of view this improvement could result from a centralization of information (or at least a minimum of information) concerning the plates stored in each plate vault. As an example, duplication of work could be avoided.

We carried out an enquiry last summer among observatories possessing a plate vault. The response rate to the questionnaire was very high (88%) and the percentage of answers in favour of centralization also very high (86% of the replies on this point). I wish to express my gratitude to all who replied to our questionnaire.

The replies we received concern more than 1.5 10^6 plates and the major problem will be the organization of a central file. Table 2 gives a summary of the replies and table 3 details of each observatory's reply. Five observatories answered favourably to the question "Would you be willing to act as 'centralizer'", three would accept also but with restrictions, and proposals were made to ask the Stellar Data Center to act in this way. We may therefore envisage the future on a solid basis. In my opinion it would be interesting to keep a master file which is regularly updated by information from all observatories and to distribute copies to three or four other locations. As at present only a few observatories have a catalogue of plates in their plate vault which is usable by computer, it would be somewhat ambitious to wish to centralize too much information. On the contrary, a minimum of information is the only way of ensuring success.

Table 2. Summary of Replies

Number of questionnaires sent: 48 Returned: 42 (88%)
32 observatories have a plate vault
2 " " plates but not an organized plate vault
2 " will have a plate vault in the near future
6 " do not have any plates
Number of plates involved: at least 1.5 10^6
Nearly all plate vaults are accessible to visitors
20 observatories are willing to loan their plates outside
8 " " " with some restrictions
32 " have a list or card file of plates (for all or only a part)
In 9 cases the file is usable by computer
31 answers are favourable to centralization of information

Table 3 - Replies to the Questionnaire

Observatory	Plate vault?	Person in charge	No. of plates	Loan of plates outside	Card file or list	File usable by computer
Australia						
Anglo-Australian Observatory P.O. Box 296, EPPING, N.S.W. 2121	Yes	D. Malin	approx. 2000	subject to some conditions	Both	card/tape
Austria						
Institut für Astronomie Türkenschanzstr. 17, 1180 WIEN	Yes	A. Schnell	~ 3100	subject to some conditions	List	No
Canada						
Mont Mégantic Observatory C.P. 6128, MONTREAL, Quebec H3C 3J7	No					
David Dunlap Observatory RICHMOND HILL, Ontario L4C 4Y6	Yes	C.T. Bolton	46000 +	Yes	Yes	No
Dominion Astrophysical Observatory 5071 West Saanich Road, VICTORIA, B.C.	Yes	E.K. Lee	110000	Yes	File	Yes
Chile						
Observatorio Interamericano de Cerro Tololo, Casilla 63-D, LA SERENA	No					
France						
Observatoire de Haute Provence Saint Michel l'Observatoire 04300 FORCALQUIER	Yes	H. Genevoix	98000	No	Yes	No
C.E.R.G.A. ST VALLIER DE THIEY	Yes	J.-L. Heudier	400 increasing	Yes	Yes	Yes
Observatoire de Toulouse 1 avenue Camille Flammarion 31500 TOULOUSE	Yes	R. Nadal	9012	Yes	2 Lists	No
Germany, Federal Republic						
European Southern Observatory GARCHING bei MÜNCHEN	Yes	R.M. West	4000	Only some	Computer list	Will be
Hamburger Sternwarte Gojenbergsweg 112, 205 HAMBURG	No					
Landessternwarte Königstuhl HEIDELBERG 1	Yes	G. Klare	15200	No	Schurig-Götz atlases + catalogues	Yes
Max-Planck-Institut für Astronomie 69 HEIDELBERG 1	Foreseen					
Germany, Democratic Republic						
Sternwarte Sonneberg, 6400 SONNEBERG	Yes	W. Wenzel	170000	No	Lists	Under consideration
Karl-Schwarzschild Observatorium 6901 TAUTENBURG	Yes	S. Marx	5500	Yes	Yes	No
Indonesia						
Bosscha Observatory, LEMBANG, Java	Yes	B. Hidayat	7000	Yes	Yes	No
Israel						
Weizmann Institute of Science, REHOVOT	No					
Italy						
Astrophysical Observatory of the University of Padova 36012 ASIAGO (Vicenza)	Yes	R. Barbon	37498 (direct phot.) 21135 (spect. & ob. prism)	Yes	List (partial card file)	Yes
Osservatorio Astronomico Universitario via Zamboni 33, 40100 BOLOGNA	Yes	F. Bonoli	19000	Not all plates	In preparation	--
Osservatorio Astronomico di Brera (Milano-Merate), via Brera 28 20121 MILANO	Yes	E. Antonello	9000	No	No	--

Table 3 - Replies to the Questionnaire (contd.)

Observatory	Plate vault?	Person in charge	No. of plates	Loan of plates outside	Card file or list	File usable by computer
Japan						
Sky Patrol Section, Tokyo Astronomical Observatory, Mitaka, TOKYO 181	Yes	Y. Kozai	5500	Yes	List	No
Kiso Branch of Tokyo Astronomical Obs.	Yes	B. Takase	2850	No	List	Tape
Okayama Astrophysical Station Tokyo Astronomical Observatory	No					
Netherlands						
Sterrewacht Leiden, Huygens Laboratorium Postbus 9513, 2300 RA LEIDEN	Yes	A.A. Schoe-maker	~ 30000	Yes	Yes	No
Poland						
Astronomical Observatory N. Copernicus University, 87-100 TORUN	Yes	A. Strobel	1600	No	Card file	No
South Africa						
South African Astronomical Observatory P.O. Box 9, OBSERVATORY 7935	Yes	Director	126000	In certain cases	Incomplete	No
Sweden						
Lund Observatory, Box 1107 22104 LUND	Yes	--	~ 3000	Some	No	No
Uppsala Astronomical Observatory Box 515, 750120 UPPSALA	Yes	Not yet named	Large	Yes	Partial	No
United Kingdom						
Royal Greenwich Observatory, Herstmonceux Castle, HAILSHAM, East Sussex BN27 1RP	Yes	E.D. Clements	90000	Yes	Both	Yes
Royal Observatory, Blackford Hill, EDINBURGH EH9 3HJ, Scotland	Yes	S.B. Tritton	5000	Yes	Yes	Yes
U.S.A.						
University of Michigan, Dept. of Astronomy ANN ARBOR, MI 48109	Yes	W. Hiltner	>8000	No	Partial	No
McDonald Observatory, University of Texas AUSTIN, TX 78712	No					
Harvard College Observatory, 60 Garden St. CAMBRIDGE, MA 02138	Yes	M. Liller J. Kloss (acting)	400000	Seldom	Card file	Yes
Lowell Observatory, P.O. Box 1269 FLAGSTAFF, Arizona 86002	Yes	A. Hoag	Many	Yes	No	No
Institute for Astronomy, University of Hawaii, 2680 Woodlawn Drive, HONOLULU	Yes	W.K. Bonsack	3500	Yes	No	No
Canada-France-Hawaii Telescope Corporation P.O. Box 1597, KAMUELA	Foreseen	Director	300	Yes	Commenced	Perhaps
Mount Wilson and Las Campanas Observatory 813 Santa Barbara St., PASADENA, CA 91101	Yes	Director	Thousands	Limited	Yes	No
Lick Observatory, University of California, SANTA CRUZ, CA 95064	Yes	G. Herbig	70000 +	Yes	Partial list	No
Kitt Peak National Observatory, P.O. Box 26732, TUCSON, Arizona 85726	Yes	A. Hewitt	~3500	Yes	List	No
Yerkes Observatory, WILLIAMS BAY, Wis 53191	Yes	K. Cudworth	~150000	Yes	Several	No
U.S.S.R.						
Abastumani Astrophysical Observatory 383762 ABASTUMANI, Georgia	Yes	D. Chipashvili	83600 +	Yes	List	No
Special Astrophysical Observatory, USSR Academy of Science, 357140 N. Arkhyz Stavropol Region	Yes	J. Glagolevskij	3300	No	Yes	No

REFERENCES

Blanco, V.M., Demers, S., Douglass, G.G., and FitzGerald P.M.: 1968, Publ. US Naval Obs. 2nd, p. 21
Hauck, B., and Mermilliod, M.: 1980, Astron. Astrophys. Suppl. 40, p. 1
Hauck, B., and Mermilliod, M.: 1981, Bull. Inf. CDS no 21
Jaschek, C.: 1978, Quart. J. Roy. Astron. Soc. 19, p. 269
Jaschek, C.: 1980, CODATA Bulletin no. 36
Jaschek, C., Coude, H., and de Sierra, A.C.: 1964, Obs. Astron. La Plata, Seria Astronomica XXVIII
Jaschek, M.: 1978, Bull. Inf. CDS 15, p. 212
Mermilliod, J.C., and Nicolet, B.: 1977, Astron. Astrophys. Suppl. 29, p. 259
Nicolet, B., and Hauck, B.: 1977, in Compilation, Critical Evaluation and Distribution of Stellar Data, C. Jaschek and G.A. Wilkins (eds.), D. Reidel, Dordrecht, p. 121
Rossmassler, S.A., and Watson, D.G.: 1981
Rufener, F.: 1981, Astron. Astrophys. Suppl. in press
Woltjer, L.: 1978, in Optical Telescopes of the Future, F. Pacini, W. Richter and R.N. Wilson (eds.) ESO Geneva, p. 5

PROPOSAL FOR CENTRALIZATION OF INFORMATION CONCERNING PLATE VAULTS

B. Hauck
Institut d'Astronomie de l'Université de Lausanne
CH-1290 Chavannes-des-Bois

The replies to the questionnaire of the Working Group on Astronomical Data (IAU Commission 5) were favourable to centralization of information concerning plate vaults. Some observatories have a file recording the plates which is usable by computer and some publish their lists regularly, for example, the Haute-Provence Observatory and the Tokyo Astronomical Observatory (Kiso Information Bulletin). Thus we have a basis to start centralizing this information. However, the most important task will be to enable other observatories to participate also in this collaboration. Bearing this in mind, I propose that the "central file" contains, at least at the beginning, only the minimum of information necessary.

I have attempted to define this minimum and propose that we keep, for the spectra, the following items:

a) object identification: HD or BD or ...
   It will also be necessary to foresee the identification of objects such as comets, planetary nebulae, small planets, etc.
b) (α, δ)
c) m_V
d) UT date and LST or Julian date
e) plate designation

For the Schmidt plates I would propose

a) (α, δ) of the centre
b) angular size
c) UT date and LST or Julian date
d) direct image or objective prism, filter
e) plate designation

This information is perhaps insufficient and could be completed by further details, for example, for the Schmidt plates the limiting

C. Jaschek and W. Heintz (eds.), Automated Data Retrieval in Astronomy, 227–228.
Copyright © 1982 by D. Reidel Publishing Company.

magnitude, emulsion and exposure time. This matter is open for discussion.

I now wish to mention briefly my opinion regarding the location of this information. It has been suggested that the Stellar Data Center could act as "centralizer" of this information. This center has undoubtedly been very successful in astronomical data retrieval and it would be a guarantee of our success if it were possible to arrange something with Strasbourg. On the other hand, some observatories answered favourably to our question and in view of their geographical location I think that the best solution would be to have the master file here at Strasbourg which would be updated regularly by information received from each observatory and copies could be held at Harvard Observatory and the Anglo-Australian Observatory. Both these observatories already have files usable by computer. This is also a matter for present discussion.

THE UK SCHMIDT TELESCOPE PLATE CATALOGUE AND PROBLEMS ASSOCIATED WITH INCREASING NUMBERS OF PLATES AND USERS

S.B. Tritton
Royal Observatory, Blackford Hill, Edinburgh EH9 3HJ.

The UK 1.2m Schmidt Telescope (an outstation of the Royal Observatory Edinburgh) is situated at Siding Spring Observatory in New South Wales, Australia. The telescope has been operational since 1973 and has, to date, taken about 7000 plates. This paper discusses some of the problems associated with cataloguing these plates and keeping track of their locations.

The main initial task of the telescope was to produce the original plates to be used in making the "blue" half of the ESO/SRC Southern Sky Survey; this task is now almost complete. The telescope is also being used to produce other sky surveys, in particular the Infrared Survey of the Southern Milky Way and an extension to the equator of the Southern Sky Survey in both blue and red wavebands. As well as taking survey plates the UKST invites applications from astronomers for special plates to be taken for their own research programmes and about 50% of plates taken are for these special programmes. Each new application is given a number and there have so far been 480 such requests of which over 200 are still active and material is still on loan to a further 100. All plates remain the property of the UKST unit and cataloguing these plates and keeping track of their locations is a major task.

Due to the lack of computing facilities at Siding Spring all cataloguing of the plates, research applications etc. is done at ROE. Each month a log of plates taken is sent to Edinburgh and the data are added to the computer catalogue. Each plate is uniquely identified by a running serial number; other data included in the catalogue include the RA and Dec of the plate centre, the date, time and length of the exposure, information on the filter and emulsion used and a grade to indicate the quality of the plate. These data have been chosen to try and help users of the catalogue select suitable existing plates for use in their research projects. Information on research applications is also stored on the computer, this includes a programme title and the names and addresses (which

C. Jaschek and W. Heintz (eds.), Automated Data Retrieval in Astronomy, 229–231.
Copyright © 1982 by D. Reidel Publishing Company.

change all too often!) of the investigators. A separate file contains data on plates which are on loan.

Keeping track of the location of the plates is particularly prone to errors (for example, the loan is entered when a plate is first loaned but not removed from the record when it is returned) a problem which is not helped by the current mobility of astronomers who may take their plates with them or simply abandon them at their previous institution. Some plates are kept in Australia, these are mainly the accepted original survey plates and plates being used by Australian based astronomers for their own research, all others are returned to ROE. Of the latter plates some will be kept in the ROE plate library for general reference, some will be kept in the plate library while actually being on loan to ROE astronomers while others will be sent to users in the UK or overseas (mainly Europe and the USA although some have gone as far afield as China).

A further complication arises when the original plate is copied, perhaps because the original is to be used for Atlas production, perhaps so that copies can be distributed to several users for use in different programmes or simply as an insurance in case the original is damaged. The details of the copies made, whether positive or negative, glass or film, and their location need to be recorded.

At present the data are all stored on the GEC 4082 computer at ROE. The basic catalogue only is also available for consultation via the Starlink network and a series of programs for interrogating the catalogue are planned to be available for general use. All existing programs work only on the GEC and are therefore only available at ROE. Most of the programs are fairly cumbersome and have been designed with the object of producing detailed and lengthy printouts. For example, it is possible to obtain a listing of all plates taken to date together with details of the present location and the current user (if any). These data rapidly become out of date as plates are returned to the library and possibly reloaned. To check the current status of a plate it is necessary to obtain a full printout of all plates taken. The data in the basic catalogue change only occasionally (and then usually due to the discovery of an error!) while the loan data change much more frequently. When there were only a few thousand plates it was not unreasonable to produce a complete printout every month. Now this task is becoming very wasteful of computer time and paper and there is a very obvious need for a system to interrogate the plate loan file independently of producing a complete listing. A further need, which is becoming more obvious every day, is to keep a complete record of a particular plate's history. It is not only necessary to know where a plate is today, it can also be necessary to know its loan history, for example when two astronomers are using one plate for different research programmes. At present, all past history is deleted from the computer records when a plate is returned to the library.

With a total of about 6000 plates (excluding test plates) the present system is only just managing to keep up with the taking and movement of plates. It is now apparent that a database management system is required to enable UKSTU to keep a full check on all the plates. Ideally, we would like the DBMS to be implemented on the Starlink network; this would be particularly useful in providing expertise in the use of the system and would also enable astronomers all over the UK to consult the catalogues with relative ease. While several of the problems specified above could be solved by writing Fortran programs, this can be both inefficient and time-consuming. A DBMS should allow most of the present problems to be solved fairly easily and would thereby free staff to implement a full interrogative package for general use. At present two staff members are involved in checking the whereabouts and loan status of all plates which have been returned to the UK, this has taken up a considerable amount of their time for several weeks and it seems unlikely that such a check will be possible in the future. It is therefore becoming very important that a more reliable computer system is introduced with as little delay as possible.

Two problems will not be solved even if we obtain a DBMS. The plate log will still be handwritten in Australia and the errors inevitably introduced due to copying the data several times will still occur. Nor will we be able to keep a full record of the plates held in Australia (at present we only note that they are in Australia). These problems will only be solved with the introduction of a similar system in Australia so that the data can be sent directly to Edinburgh in computer readable form. With luck, in a year or two we will have a fully working DBMS in Edinburgh and a suitable computer in Australia and will then be able to offer a much better service to the several hundred users of plates taken by the UK Schmidt telescope.

RETRIEVAL OF ASTRONOMICAL INFORMATION FROM PADOVA-ASIAGO OBSERVATORY PLATES ARCHIVES

Leopoldo Benacchio
Osservatorio Astronomico di Padova

This paper shortly describes a software procedure for the on line information retrieval from astronomical plates archives. The procedure is active in Padova Asiago Observatory and will be improved linking them with other retrieval and graphic procedures acting on astronomical catalogues.

1. BACKGROUND

The information retrieval from astronomical plates archives is the first step to extract information from a single plate contained in the archives itselves. This is evident, for example, in tasks such a sky surveys for Nova, Supernova (SN), or variable stars. When a SN is detected, the retrieval of plates that contain the same sky area of the SN is the first task to be done. In a more general way, when the number of plates of the archives is large, a synchronous view of all the useful pieces of information becomes quite impossible without using a software procedure that allows to retain all the records of the archives plates and to search through them the desired plates, if they exist.

The plates contained in the Padova Asiago Observatory archives are those produced with the four telescopes that operate in Asiago. In the table at the top of the next page the more important figures of the main plate "types"are shown. In the observations log books a single plate is identified by two parameters : the name of the register, that is the name of the telescope used and the serial number inside this register. Each record consists of entries common to all types of observation : date, α , δ , observer name, emulsion type, filter, sky conditions, general remarks, and of some other observations and "boundary" conditions (type of spectrograph, slit conditions, dispersion used, type of intensifier).

C. Jaschek and W. Heintz (eds.), Automated Data Retrieval in Astronomy, 233–236.
Copyright © 1982 by D. Reidel Publishing Company.

ASIAGO TELESCOPES

	Type	N° of Plates	Operating from
122 CM	Photo	9492	1942
	Spectra	16753	1949
50/40 Schmidt	Photo	14498	1957
	Obj. Prisme	1862	1962
90/60 Schmidt	Photo	10751	1964
	Obj. Prisme	580	1966
182 CM	Photo	2757	1972
	Spectra	1940	1973
		58663 as at 31.12.1980	

2. SOME PROBLEMS

There are some problems in switching such a kind of register forms from paper to magnetic support. The most important is that a standardisation of the information record concerning an astronomical information does not exist until now. A second problem is : how many pieces of information can be considered common to all observations and how many are sufficient to retain a specific type of observation ? A third one that is coming up is that the record of new types of images can be completely different from that of the plates (i.e. Reticon or Digicon images, CCD or satellites images). It is easy to forecast that the number of these images will grow quickly ; also for these images it is very important to define right now a standard record.

3. THE PROCEDURE IN PADOVA

A procedure for the on-line retrieval of astronomical information from the data base built with the information registers exists at the Padova Observatory. All the information contained in the paper registers has been copied on disc files in order to be able, in the future, to fit a standard different from that in use in Asiago. The data base consists of eight files containing the eight observation log books. The name of the file identifies the telescope used and the type of observation. The serial number inside the register directly defines the record position inside the file ; in this manner a plate is completely identified, in the disc volume, by these two parameters with which astronomers are generally familiar. Each plate record in the data base consists of eight different entries common to all observations (object name, α , δ , plate type, filter, exposure, year, date) and of a "comment-field" where specific conditions or remarks are stored (type of spectrograph, intensifiers, slit condition, etc...)

A block diagram of the procedure is sketched in fig. 1. A main program manages the facilities offered by the procedure ; search, display and print (lists, graphics). An Help program can be invoked by the user after each question posed by the procedure. It gives a two level information : the first is a longer description of the question and of the possible answers, the second is represented from the access to the procedure user manual that is loaded on line. The search program has two different branches : one to search only in Schmidt plates files if a defined area in the sky has been covered in the past, another, more general, to make a sequential search in all the entries of the data base. The user at the start of the session can select from one to eight files, that means the complete data base or only a subset. Limitations can be posed to the eight entries and also to the "comment" field. In the alphabetical entries a variable length label can be searched (in fig. 2 the character @ defines the end of the label to be searched). This feature is useful in particular to retrieve labels whose codification is not well known to the user, or to retrieve all objects of a certain type (i.e. all the NGC objects). Conditions to numerical entries can be set by giving the value of an interval, including zero length. Through the use of the logical operators OR and AND the desired plates subset can be built with any combination of entries values. In the short example shown in fig. 2 conditions are posed to observer name (OB) AND to object name (OJ) AND to plate type (PT). The search is done on the photo log. book of the 182 cm telescope. An users manual is available that the contains a longer description of the capabilities of the procedure and provides examples.

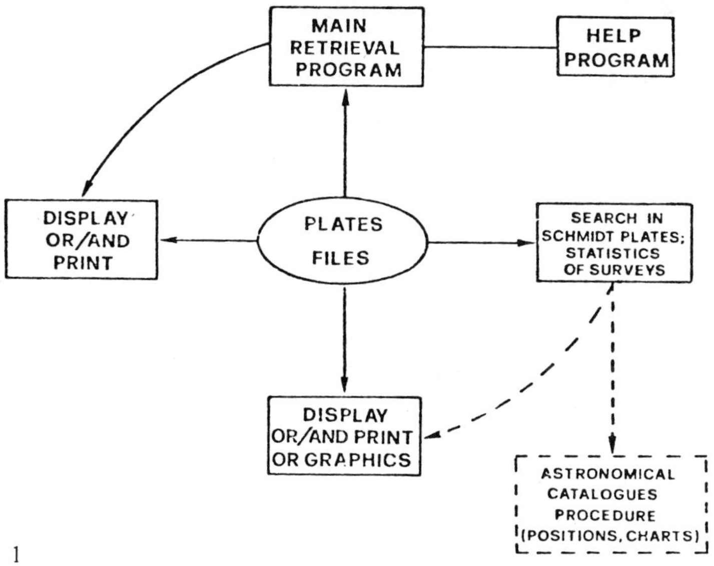

Figure 1

```
Welcom to  IRA  : Type your name and then press RETURN

DEMO

O.K. Date :11:50 AM  WED.,  1  JULY, 1981 User : DEMO
Start session
Remember  that if You answer HELP to any question from  IRA
You will have a longer description of the question
Now press RETURN to continue


+ + + + + + + Select Register ( /E To End)
T182FO
/E
+ + + + + + + Enter Key to select
OB
+ + + + + + + Input label to search ( Remember @ )
ROSN@
+ + + + + + + Finding label :ROSN
+ + + + + + + Data base in use : T182FO
+ + + + + + + Total number of plates sorted :   484

+ + + + + + + Plates sorted in this cycle /   484

+ + + + + + + Enter action : AND     OR     PRINT     EXIT   GRAPHICS
AND
+ + + + + + + Enter Key to select
OJ
+ + + + + + + Input label to search ( Remember @ )
ORION@
+ + + + + + + Finding label :ORION
+ + + + + + + Data base in use : T182FO
+ + + + + + + Total number of plates sorted :    14

+ + + + + + + Plates sorted in this cycle /    72

+ + + + + + + Enter action : AND     OR     PRINT     EXIT   GRAPHICS
AND
+ + + + + + + Enter Key to select
PT
+ + + + + + + Input label to search ( Remember @ )
103AO@
+ + + + + + + Finding label :103AO
+ + + + + + + Data base in use : T182FO
+ + + + + + + Total number of plates sorted :     2

+ + + + + + + Plates sorted in this cycle /  1242

+ + + + + + + Enter action : AND     OR     PRINT     EXIT   GRAPHICS
PR
++++++++++++++++   REGISTRO OSSERVAZIONI : T182FO     ++++++++++++++
OGGETTO : ORIONENEBULO ALFA :  $$ $$ $$  DELTA :  $$ $$ $$ LASTRA NO. :     1
DATA :   8  3  1973    T.U. :   0  0  0  POSA  :   5.00    FILTRO     :

TIPO LASTRA  : 103AO    OSSERVATORE : ROSN
COMMENTO : SERENO.LUNA ULTIMO QUARTO. NON GUIDATA

++++++++++++++++++++++++++++++++++++++++++++++++++++++++++++++++++++++++++++


++++++++++++++++   REGISTRO OSSERVAZIONI : T182FO     ++++++++++++++
OGGETTO : ORIONETRAPEZ ALFA :  $$ $$ $$  DELTA :  $$ $$ $$ LASTRA NO. :   104
DATA :   4 10  1973    T.U. :  25 42  0  POSA  :  30.00    FILTRO     :

TIPO LASTRA  : 103AO    OSSERVATORE : ROSN
COMMENTO : SERENO

++++++++++++++++++++++++++++++++++++++++++++++++++++++++++++++++++++++++++++
```

Figure 2

LARGE SCALE PHOTOMETRIC SURVEYS USING ARCHIVAL PLATES

Kenneth Janes
Boston University

ABSTRACT

Modern microdensitometers and image processing software suggest the possibility that it may soon be feasible to utilize the enormous amount of photometric and positional data potentially available on existing photographic plates. This paper describes the possibility of large-scale surveys and outlines a method for densitometric calibration of plates from an analysis of the detailed structure of the star images on the plate.

1. THE PROBLEM OF PHOTOGRAPHIC PHOTOMETRY

A photograph is an amazingly efficient medium for storing information. Each of the many thousands of star images on a plate can be located within a fraction of a millimeter and its brightness measured with a precision of a few hundredths of a magnitude. Often, however, only a tiny fraction of the stars are ever measured - it is simply too tedious to measure them all. Photography has, in fact, two major faults. Although information is stored efficiently, it is not easily retrievable (in a quantitative sense) and there is a non-linear relation between incident light and developed photographic density.

It now appears that the resolution of the first problem may be at hand. With computer-controlled microdensitometers such as the widely-used PDS machine, the density information on an entire plate can be digitized and stored on magnetic tape in a few hours. While this is still longer than it probably took to produce the photograph, it makes it possible to conceive, at least, of extracting all the information on substantial numbers of plates. In the immediate future a further improvement in speed of more than an order of magnitude seems possible using solid state array detectors to scan the plates. I have made a few tests with a scanning microdensitometer which uses a linear diode array as a detector. The array scans the scene in one direction and is stepped in the perpendicular direction to develop a

C. Jaschek and W. Heintz (eds.), Automated Data Retrieval in Astronomy, 237–239.
Copyright © 1982 by D. Reidel Publishing Company.

raster scan of the entire photograph. With this particular device, it is possible to scan in excess of 10^4 pixels/sec, much faster than a PDS machine, with its single photomultiplier tube. The chief problem at the moment is scattered light and lack of dynamic range. The manufacturer of this particular device (Eikonix, Inc. of Bedford, Massachusetts) designed it primarily to scan objects in reflected light (such as maps, etc.) and so was not particularly concerned with either problem. Swaans (1979) considered the scattered light problem in a similar device and decided that it was solvable and the dynamic range can be incresed by better electronics, possibly combined with cooling the array.

Even a PDS machine generates numbers at a great rate,making the calibration and reduction problem acute. A number of people have developed methods for obtaining magnitudes from raster-scanned star images (see Hoag, 1978 for a review). Most of these procedures were developed with the idea in mind of being able to measure large numbers of stars more-or-less automatically. In all cases, however, it has been necessary either to use plates calibrated with sensitometer spots or to use a photoelectric calibration sequence covering the same range in magnitude as the unknown stars. Unfortunately, it is often the case that the plates extend several magnitudes fainter than available standards.

2. CALIBRATION BASED ON THE SHAPES OF STELLAR IMAGES

Is it possible to develop a procedure to use the information in the images of relatively bright, photoelectrically measured, star images to determine a densitometric calibration of a plate? From some preliminary studies I have been making, it appears that the answer to this question is yes, so long a certain conditions are met. It is necessary to assume, first of all, that the density in any point of an image depends only on the intensity of the light striking that point. It is necessary to neglect, therefore, various development effects or scattered light in the measuring optics. It is also necessary to assume that the distribution of intensity within a star image (the point spread function) is the same across the entire region to be measured. A plate with comatic images could not be used in this type of analysis. If these conditions are met, then there will be some function relating density and exposure:

$$D(x,y) = f\{E(x,y)\}. \qquad (1)$$

This is of course the characteristic curve. The exposure at some point in the image is given by a relation of the form:

$$E(x,y) = P(x,y) \cdot L_* + B \qquad (2)$$

where $P(x,y)$ is the point spread function, L_* is a measure of the brightness of the star and B is the background intensity.

For simplicity, I have assumed that P(x,y) is a gaussian of unit height. With this assumption, the width of the gaussian can be found by measuring the radius of the images of several standard stars at some arbitrary density. The quantity L_* is defined by the known magnitudes of the standards, and if the background level is uniform in the region of the standards, then B can be set equal to zero. Thus equation 2 can be used to determine point-by-point the value of E(x,y), which together with the corresponding values of D(x,y) lead to a solution of the characteristic curve (eqn. 1). When the images of the fainter stars are transformed into exposure, their magnitudes can be measured.

I have tested this procedure on a set of plates of the open cluster NGC 609, with satisfying results. The r.m.s. magnitude difference from plate to plate is about ±0.04 for stars brighter than about one magnitude above the plate limit. Since this field is one of many where the standards do not reach the photographic limit, it is not possible from this sample to say whether there may be systematic deviations in the magnitudes, but the good agreement between plates with different exposures looks promising. It will be necessary to try a field with a wide range of photoelectrically determined magnitudes to see how reliable the method will be. If the initial success is confirmed, then a large number of existing plates (particularly of star clusters can be re-measured to fainter limits and with many more stars.

REFERENCES

Hoag, A.A. 1978, in Modern Techniques in Astronomical Photography, Ed. RM West and J.L. Heudier (ESO: Geneva), p121.

Swaans, L 1979, in Image Processing in Astronomy, Ed. G. Sedmak, M. Capaccioli and R.J. Allen (Trieste: Osservatorio Astronomico), p. 77.

I. DATA IN SPACE ASTRONOMY

DATA AND DATA RETRIEVAL IN SPACE ASTRONOMY

J. I. Vette
National Space Science Data Center
NASA/Goddard Space Flight Center

ABSTRACT

To illustrate the range of storage and retrieval facilities and data products that are involved in space astronomy, the present situation on data from some instruments on the following missions is presented: Small Astronomy Satellite-B, Orbiting Astronomical Observatory 3, High Energy Astrophysics Observatory 2, International Ultraviolet Explorer, Viking Orbiter, and Solar Maximum Mission. In addition the Coordinated Data Analysis Workshops conducted at the National Space Science Data Center are outlined to demonstrate the usefulness of building a problem-oriented on-line data base from instruments flown on a number of spacecraft and operating from ground-based facilities.

1. INTRODUCTION

With the advent of artificial satellites carrying instruments above the Earth's atmosphere, the spectral range of observation available to astronomers was increased enormously. The major thrusts have been in the ultraviolet (UV), extreme UV (EUV), X-ray, and γ-ray regions. Not only have a variety of instruments been flown which are sensitive in this range of the electromagnetic spectrum, but there are a variety of ways in which such data have been processed and archived for use by astronomers for further study. Several of these ways will be outlined in this introduction and more details for a specific example of each will be given. This should provide the reader with a reasonable understanding of both the data and their accessibility, or retrieval, in space astronomy at the present time.

The first mode used in conducting a flight astronomical investigation was the Principal Investigator (PI) mode, where the PI was responsible for the instrument design, construction and

C. Jaschek and W. Heintz (eds.), Automated Data Retrieval in Astronomy, 243–252.
Copyright © 1982 by D. Reidel Publishing Company.

calibration as well as the extensive data processing and analysis that follows the receipt of the telemetered data from the satellite. The example selected to detail this approach is the second Small Astronomy Satellite (SAS-B) launched in November 1972, which explored for seven months the spectral range between 30 and 200 MeV. Another approach has been the sharing of available observing time by astronomers as Guest Investigators (GI) on an instrument, which was handled in all respects by a PI. The GIs were chosen from a set of proposed tasks by a peer review process which prioritized these proposals on their scientific merit. An example of this mode of operation is given by the UV Telescope covering the 95-320 nm range on the third Orbiting Astronomical Observatory (OAO 3), which operated from August 21, 1972, until shut down on February 15, 1981.

A third approach has been used on the second High Energy Astrophysics Observatory (HEAO-2), where an orbiting X-ray telescope facility was constructed under the direction of a PI and a number of focal plane instruments were each supplied by experts, who formed members of a consortium. A Guest Observer (GO) program has also been conducted by the PI so that observations were specified by the PI, the three consortium members, or the GOs. The spectral range from 0.2-4 keV was covered by this facility. A more classical approach for astronomers is used in the case of the International Ultraviolet Explorer (IUE). A facility employing a Ritchey-Chrétien telescope performing in the region from 120-325 nm is available only to GIs, who make proposals for specific observations and are selected in a peer review process. One third of the operating time is scheduled by the UK and ESA experimenters during which the spacecraft is handled by the ESA VILSPA control center in Villafranca, Spain. The other 16 hours are controlled by Goddard Space Flight Center (GSFC) for US and non-European investigators.

A fifth variant was exemplified by the Viking Orbiter Imaging Investigation. In this case, a team of scientists specified the characteristics of twin TV cameras operating in the visible spectrum and the photographic products which would be provided to them. This team consisted of 14 scientists from nine institutions. Although there was a proprietary data rights period for this team, other scientists were also supported to carry out research utilizing the photographic data products.

The last approach that will be detailed here is that utilized by the Solar Maximum Mission (SSM). In this mission, each of seven instruments was contributed by a PI-led team (in one case there were three Co-PIs involved). A very tight command and control loop with five of the instruments, which has required separate computer systems for each operating in near real-time, has been conducted. The detailed analysis of the data will be carried out in a coordinated fashion on a single computer, which is called the Data Analysis Center (DAC). The

SMM solar pointing platform is instrumented to make observations from the IR through the γ-ray region.

For these specific space astronomy projects, the details of the measurements, the data products available, and the sources for such products will be given. For most of the missions, data are deposited in the National Space Science Data Center (NSSDC) at GSFC. Any data deposited at NSSDC are available to scientists outside the United States through the co-located World Data Center A for Rockets and Satellites (WDC-A-R&S). For brevity in what follows, the acronym NSSDC will be used to represent NSSDC/WDC-A-R&S.

2. DATA FROM THE SAS-B GAMMA-RAY TELESCOPE

The instrument consisted of two spark-chamber assemblies, four plastic scintillation counters, four Cerenkov counters, and an anti-coincidence scintillation counter dome assembled to form a telescope. The spark-chamber assembly consisted of 16 wire modules with thin tungsten plates interleaved with the planes of wires to serve as electron-positron pair converters. The measurement of the resulting charged pairs allowed the determination of the direction and energy of the incident γ-ray. The four scintillation counters formed a plane between the two spark-chamber assemblies. The four directional Cerenkov counters, placed below the assemblies, constituted four independent counter coincidence systems. The single-piece plastic scintillator dome surrounded the whole assembly, except at the bottom, to discriminate against viewing charged particles passing through the chamber. The angular resolution of the telescope was about 2.5° although it varied with energy and angle. The dome scintillator was sensitive to γ-ray bursts with energies in the 150-600 keV range.

The processing and analysis of these data were carried out under the direction of Dr. Carl E. Fichtel, GSFC. A number of Co-Investigators (CoIs) were involved with these activities. Approximately three years after launch the following data products were submitted to NSSDC for archiving. There is a microfilm data set containing pictures of the electron pair spark-chamber tracks for over 3 million γ-rays. In connection with γ-ray burst activity there are tables, plots and magnetic tapes containing the dome anti-coincidence counter rate as a function of time. As a final product, there are hard copy tables of γ-rays observed in the energy band 35-100 MeV and $>$ 100 MeV in each of 144 elements of equal solid angle with a 2.5° latitude bin. The exposure factor for each element is also provided. During the lifetime of the experiment from November 15, 1972, to June 8, 1973, approximately 55 percent of the celestial sphere, including the galactic plane was surveyed. Any other data still existing from this experiment are held by the PI.

3. DATA FROM THE OAO 3 UV TELESCOPE

The instrument consisted of a 0.8-m diameter Cassegrain telescope with a 16-m focal length (f/20). A Paschen-Runge spectrometer with two pairs of movable photomultiplier tubes (PM) provided both high and low resolution spectra in ranges 95-145 nm and 160-320 nm. The high resolution was 5 and 20 pm; the low resolution was 10 and 40 pm, respectively. The high resolution made it impractical to obtain complete spectral coverage of every object observed. Stabilization of the whole device was about 0.1" in pitch and yaw. Professor L. Spitzer of Princeton University was the PI for this investigation. The GIs accounted for about 20 percent of the observations that were made. The amount of processing accomplished by the PI for the GI was small. The raw data consisted of encoder positions and PM counts. The processing consisted of converting the encoder positions into temperature uncorrected wavelength values in the star's reference frame and the PM counts into source plus background photons. In addition, a background parameter was provided. This was the form of data provided to the GIs. The temperature correction and the removal of background for the short wavelength PM tubes were trivial. However background corrections for the long wavelength were rather complex. All analysis routines were supplied by the individual GIs at their own facility.

The PI was responsible for all requests for data until the completion of the program. The data began being archived in the form provided to the GIs at NSSDC several years ago. A copy of the algorithms that are necessary to perform the final reductions will soon be provided. In addition, an observing catalog with extensive cross references will be provided so that all the observations of a single source can be located and reduced to final form. For a few objects complete low resolution spectral scans were made. NSSDC has on magnetic tape a spectral atlas for Tau Scorpii and for Iota Herculis.

4. DATA FROM HEAO 2

The Einstein Observatory, HEAO 2, had a large grazing incidence X-ray telescope with a 1° x 1° field-of-view (FOV). There were four interchangeable focal plane detector systems. The first was a channel plate imaging array with a pixel size of about 2" sensitive in the range 0.2-4.0 keV. This system and the telescope were the responsibility of Professor R. Giacconi and Dr. H. D. Tananbaum of the Harvard College Observatory/Smithsonian Astrophysical Observatory Center for Astrophysics (CFA). A bent-crystal Bragg spectrometer employing six different crystals was provided by Prof. R. Novick of Columbia University to measure X-ray line emission from 0.18 - 3 keV. The resolution, $\gamma/\Delta\gamma$, varied from the range of 50-100 up to 200-1000 depending on the crystal. A thin-window position sensitive proportional counter was used as the detector. An imaging proportional counter with a 1" resolution was used in the range 0.1 to 4 keV to study weak sources, to determine the angular structure of extended sources, and to locate objects with poorly known positions. This device was supplied by Prof. G. W. Clark of MIT. A cooled solid-state spectrometer was supplied by Dr. S. S. Holt of GSFC. A lithium-drifted detector was operated at 120° K and spectral measurements between 0.5 and 4 keV were done with a resolution of 120-150 eV full-width at half maximum.

Prof. Giacconi was the PI for all detector systems and all observations were processed at CFA to correct for attitude, merge data, search for the existence of sources, determine source location and record spectral strengths. The algorithms to process the various detection systems were supplied by consortium members.

Observation time was divided among the PI, consortium members, and GOs. The GO observing time was 20 percent during the first 18 months of the program and built up to about 35 percent by the end of mission life. The spacecraft was launched November 13, 1978 and operated until April 25, 1981. Production tapes were copied and sent to MIT, GSFC, and Columbia for the consortium observations. Production files are maintained on magnetic tape; summary files are printed and microfiched for storage. About three months after each observation was made, all data had been processed. At that time a GO would go to CFA to analyze the data on the Einstein computer system. Analysis programs to handle spectra and time history as well as smoothing and deconvolution routines, were available for use by the GOs.

The GO program will continue for several years with the PI handling the data. An observing catalog (OCA) which gives the scheduled target name, coordinates, focal plane instrument, and observing interval through February 28, 1981, has been deposited with NSSDC. This OCA will be updated to include the whole mission lifetime by November 1981. A year later the OCA will be updated to include actual durations and dates, as well as an estimate of the time during which good data were accumulated for each target. At that time a preliminary catalog of detected sources will be submitted. Finally, a source catalog and spectral information for all observations will be deposited in NSSDC by September 1984.

5. DATA FROM IUE

The IUE telescope has a 45-cm beryllium primary and a 9-cm silica secondary mirror. The instrument is 1.30 m long, has a 6.75-m focal length (f/15) and a 16'-FOV. There are two Echelle spectrographs that use SEC Vidicon cameras. Each spectrograph has a high resolution mode of 20 pm and a low resolution mode of 0.6 nm. One covers the range

113.4-208.5 nm in low and 119.2-192.4 nm in high resolution. The other covers the range 180-325.5 nm and 189.3-303.1 nm in low and high resolution, respectively. A circular entrance aperture of 3" is always open and, additionally, a 10" x 20" slit can be used.

The GOs chosen by NASA come to the IUE facility at GSFC when their observations are being made so they can determine the actual exposure times needed on each source by examining the data in near real-time. All observations are processed by the project (US or ESA) to produce geometrically and photometrically corrected spectra on tape, as well as to produce a photographic output that displays the raw spectrum and the processed spectrum in image format. This is known as the photowrite. A line plot of the observed spectrum is also provided to the GO. Any further processing or analysis is done by the GO at facilities to which he has access. The same procedure is used at VILSPA for the European GOs.

Approximately six months after the standard output products have been given to the GO, data on magnetic tape are available at three data centers: WDC-C-R&S in the United Kingdom, VILSPA, and NSSDC. Data are also available from NSSDC on a proprietary basis via special request forms signed by the GO, who thus releases his data to a requesting colleague prior to the standard release date. For observations made from VILSPA there are no photowrites archived. Photowrites of observations made from GSFC are available from NSSDC. IUE was launched January 27, 1978 and is expected to be operational for many more years.

6. DATA FROM THE VIKING ORBITER IMAGING INVESTIGATION

The Viking visual imaging subsystem consisted of twin high-resolution, slow-scan television framing cameras mounted on the scan platform of each orbiter with the optical axes offset by 1.38°. Each camera consisted of a 475-mm focal length telescope, a 37-mm diameter Vidicon, the central section of which was scanned in a raster format of 1056 lines by 1182 pixels, and six color filters to restrict the spectral bandpass of an image to limited portions of the camera's response. The FOV was 1.54° x 1.69° with each pixel subtending 25 radians. The data were processed by the Imaging Processing Lab (IPL) at the Jet Propulsion Lab (JPL) to produce a digital image tape and the supplementary experiment data records (SEDR) which contain 78 parameters associated with each scene. Besides the tapes, the members of the Imaging Science Team headed by Dr. M. H. Carr of the US Geological Survey (USGS) at Flagstaff received processed photographs, both rectilinear and orthographic, of each scene. Additional processing was done by team members and mosaics were made from various photographs. The archive for the image tapes is at JPL. The following photographic products are available from NSSDC:

a. Press Release Photos, Black/White (B/W) & Color 4"x5"
b. B/W Rectilinear Photos 5"x5"
c. B/W Orthographic Photos 5"x5"
d. B/W Mosaics
e. Indexes on microfilm & microfiche
f. SEDRs

All photographic products can be viewed at NASA Regional Planetary Image Facilities (PIF) located at: USGS in Flagstaff, AZ; U. of Arizona in Tucson, AZ; JPL in Pasadena, CA; Washington U. in St. Louis, MO; Cornell U. in Ithaca, NY; Brown U. in Providence, RI; and the Lunar and Planetary Institute in Houston, TX. In addition, cooperative PIFs located at U. of Rome; Rome Italy and U. of London, London, UK also have the photographic products. The Viking 1 Orbiter began operations around Mars on June 19, 1976, and terminated on August 7, 1980. The Viking 2 Orbiter dates were August 7, 1976, and July 25, 1978, respectively. NASA is in the process of generating 35-mm images from all of the planetary image tapes at JPL, which will then be used to produce video disk masters containing 54K images per side. Copies of these disks will be provided to the NASA Regional and Cooperative RIFs as well as NSSDC.

7. DATA FROM SMM

There are seven separate experiments on SMM. Each will be discussed briefly since the data from at least six are going to be resident at the SMM DAC at GSFC. This is a VAX 11/750 computer facility that is being formed to analyze the data from SMM in a coordinated fashion. There has been and will continue to be a GI program. NSSDC will have access to the DAC and appropriately identified data in the DAC will be available to other scientists through NSSDC. Eventually all SMM data processed to the proper level and documented will be archived at NSSDC.

7.1 Hard X-ray Imaging Spectrometer (HXIS)

This instrument (PI - Prof. C. de Jager, U. of Utrecht) produced two-dimensional images with 8" resolution over an approximately square area of 2' 40" per side and with 32" resolution over an approximately square area of 6' 24" per side. These images were observed in six selectable energy channels between 3.5 and 30 keV with a temporal resolution of 0.5-7s, depending on the mode of operation. By means of a flare flag the experiment alerted all the other SMM instruments when a flare began and indicated the position of the brightest pixel. The instrument consisted of 10 etched grid plates, each divided into 576 sections that formed the collimator and 900 mini-proportional counters that provided the detector and spectral capability. A dual microcomputer system permitted three modes of operation with commandable parameters that provided a flexible trade-off between temporal and spatial coverage during different phases of a solar flare.

7.2 Soft X-ray Polychromator (XRP)

This instrument (Co-PIs: Dr. L. W. Acton, Lockheed Palo Alto Lab; Dr. J. L. Culhane, Mullard Space Science Lab; Dr. A. H. Gabriel, Appleton Lab) consisted of two Bragg crystal spectrometers. A bent-crystal spectrometer using eight crystals and position-sensitive proportional counters covered seven bands from 0.177-195 nm for Fe lines and 0.317-0.323 nm for the CA XIX line with a 6' x 6' FOV. A flat-crystal spectrometer using seven crystals and proportional counters was capable of scanning the range 0.14-2.24 nm and rastering in 5" steps over a 7' x 7'-FOV with 14" resolution. A microprocessor controlled trade-offs between temporal and spatial resolution with a temporal limit of 64 ms.

7.3 UV Spectrometer and Polarimeter (UVSP)

This instrument (PI - Dr. E. A. Tandberg-Hanssen, Marshall Space Flight Center) consisted of a Gregorian telescope, an Ebert grating spectrometer and a polarimeter that used four CsI PMs and one CsTe PM. The telescope had a 1.8-m focal length with a 66.4-cm^2 aperture and a FOV of 256" x 256". Entrance slits of 1" x 1", 3" x 3", 10" x 10", 30" x 30", 1" x 10", 4" x 4", 15" x 286", and 1" x 180" were available. Rastering over any of the FOV in 1" x 1" steps was possible. The grating was 3600 lines/mm with a FL of 1 m. The spectral range covered was 115-180 nm in 2nd order and 175-360 nm in 1st with spectral resolutions of 2 pm and 4 pm, respectively. The polarizer consisted of two MgF waveplates with 22.5° rotation steps. A microprocessor made this an extremely versatile instrument that could be programmed into a number of sequential modes that could be altered based on the output of HXIS.

7.4 Hard X-ray Burst Spectrometer (HXRBS)

This instrument (PI - K. J. Frost, GSFC) observed the full solar disk in the range 20-260 keV in 15 differential energy-loss channels with a temporal resolution of 128 ms. It consisted of a disk-shaped CsI (Na) central crystal and a CsI (Na) active collimator element that surrounded the central crystal and provided a 40° conical FOV. A circulating memory was also used to provide 1-ms resolution for fast rising bursts without spectral information. Either a constant time or constant count mode could be selected for use with the memory. Using the latter mode, γ-ray bursts could be detected during spacecraft night.

7.5 Gamma-Ray Spectrometer (GRE)

This instrument (PI - Prof. E. L. Chupp, U. of NH) utilized a set of NaI (Tl) and Cs I (Na) crystals to form three separate devices: (1) an actively shielded multi-crystal γ-ray spectrometer, (2) a high-energy γ-ray detector, and (3) an auxiliary X-ray detector. The spectrometer produced a 476-channel energy-loss spectrum every 16 s over the range 0.3-9 MeV. A 2-s time resolution was available in three windows in the 3.5-6.5 MeV range to study prompt line emission at 4.4 and 6.1 MeV. Photons from 0.3-0.35 MeV were recorded with 64-ms resolution. Events in the 10-100 MeV photon range could be observed along with neutrons > 20 MeV. The auxiliary X-ray detector made measurements in the 10-80 and 25-140 keV range with 1 s resolution.

7.6 Corongraph/Polarimeter (C/P)

This instrument (PI - Dr. L. House, High Altitude Observatory) consisted of a coronograph, a polarimeter, and an SEC Vidicon detector. The FOV ranged from 1.6-6 solar radii (SR) in sectors from 1.5 x 1.5 to 6 x 6 SR with a pixel resolution of 10" and size of 6.4". Seven filters covering the 444.8-641.8-nm range were used. Three polaroids with axes 60° apart were used.

7.7 Active Cavity Radiometer Type IV (ACRIV)

This instrument (PI - R. C. Willson, JPL) used three independent pyroheliometers with conical-cavity detectors. The total solar emission from the EUV to the far IR was measured. Each device had a shutter to block the solar radiation and a reference cavity that permanently viewed a surface at known temperature.

7.8 SMM Data Processing

The five experiments HXIS, XRP, UVSP, HXRBS, and C/P had computers of the PDP 11/34 class at GSFC in a facility called the Experimenters Operation Facility (EOF) to process data within 24 hours to plan the next days' schedule. False color images from all but C/P and HXRBS were produced at the EOF. The GRE and ACRIV data were processed at the PIs' home facilities. In addition, false color images and electron density profiles from C/P were processed at the PI's home computers.

7.9 Status of SMM Operations

The satellite was launched on February 19, 1980. By September 23, 1980, a sequencer on C/P malfunctioned ceasing scientific data collection. By November 23, 1980, the spacecraft could only point within 10° of the sun so all experimenters except HXRBS, GRE and ACRIV ceased taking data. The wider FOV instruments are still acquiring data.

8. COORDINATED DATA ANALYSIS WORKSHOPS (CDAWs)

In viewing the archiving and retrieval of space astronomy data, it should be apparent that the main body of this data base resides mainly on magnetic tapes, microform (film and fiche), and photographic film. Small portions at any given time are resident on disks at computer facilities such as the Einstein computer at CFA and the SMM DAC. In association with the fields and particles experimenters involved in solar-terrestrial research, NSSDC has developed a process called CDAW in which problem-oriented data bases have been constructed from instruments flown on 8-10 satellites and from about 60 ground-based instruments. These coordinated data bases (CDB) contain approximately 600 physical parameters and have been limited to about 150 Megabytes. At scheduled times of 3-4 days duration a CDB has been placed on-line on the NSSDC computer and some 30 scientists from around the world have convened at NSSDC to be interactive with each other and with the CDB. In such workshops, the physical parameters have been manipulated and plotted to the specifications of the workshop participants with a typical daily output of 150 line plots.

Following such CDAWs, outputs have been requested by mail, telex or telephone from the CDB by the CDAW participants. A few participants have exercised a CDB remotely from terminals at their own institution. Portions of one of the four CDBs that have been built to date have been placed on a US-based, commercial computer network that can be accessed from Europe and Japan.

Of interest to astronomers may be a CDB that NSSDC has constructed for possible use to study the solar flare outputs on June 7, 21, and 29, 1980, detected by a variety of instruments. At present there are only a total of 35 parameters from the following five experiments: (1) X-ray monitor from GOES 2, (2) transient γ-ray sources from Pioneer Venus 1, (3) HXRBS from SMM, (4) HXIS from SMM, and (5) UVSP from SMM. Experimenters on ISEE 3 and IMP-J have also expressed interest in contributing to this CDB, known as CDB 5.

It is believed that the process of building CDBs and conducting CDAWs may be useful to astronomers in examining sources or regions from the radio frequency to the γ-ray region of the spectrum. Access to, manipulation of, and outputs from such CDBs can be available to astronomers with present day computer and communications technology.

THE "CATALOGUE OF STELLAR GROUPS" : A PROGRESS REPORT

Egret D. and Jaschek M.
Centre de Données Stellaires - Strasbourg

ABSTRACT

The catalog will provide lists of stars for about fifthy groups of determined spectral peculiarity. The adopted selection procedure, the connection to the C.D.S. data base and the future developments are described.

INTRODUCTION

The Catalog of Stellar Groups (C.G.S. : "Catalogue des Groupes Stellaires"), started in 1978, provides lists of stars about fifthy different groups with determined spectral peculiarity, like Ap, Am, H and K lines in emission, etc... The aim is to make available extensive lists of peculiar stars, especially for observational programs and statistical studies.

THE PROCEDURE FOLLOWED TO CONSTRUCT THE CATALOGUE

The procedure can be divided into six steps.

1.1. The first is the adoption of a definition for each group. This is an important point since the same denomination may have different meanings according to different authors but only one can be retained. As an example, our adopted definition of Be stars is somewhat restrictive because we accept only stars for which an MK classification is known. It is to be noted that the definitions we used are generally spectroscopic definitions : we do not attempt to include stars on the basis of their photometric properties (with the exception of the photometric variable groups, like δ Scu and β Cep). Only a few groups are defined kinematically (Pop. II objects) or by other characteristics (spectroscopic binaries).

1.2. The second step consists of a literature search for the lists of

C. Jaschek and W. Heintz (eds.), Automated Data Retrieval in Astronomy, 253–256.
Copyright © 1982 by D. Reidel Publishing Company.

members of each group. No critical editing was performed, since this is impossible if one has not seen the spectrum oneself. Lists were included if the authors used a definition of membership compatible with ours.

We started with the most comprehensive lists (or catalogues) published by the specialists, and we completed them with the more recent paper. Presently 180 references are used including general catalogues such as the Michigan Spectral Survey (Houk and Cowley, 1975 ; Houk 1978) or the list of southern stars of astrophysical interest by Bidelman and Mac Connell (1973).

1.3. For each list retained, the following data were keypunched : an identification (H.D. number when available, Durchmusterungen or coordinates when no common identificator is provided) and a code number of the bibliographic source. In some cases complementary spectral information has been punched, but this has not been made systematically. At the present time (July 81) 30 000 cards have been punched.

1.4. It happens frequently that a star appears in different references. This question is treated by sorting each file and merging records with common identifications. In consequence each star may have several source numbers. These multiple references provide sometimes conflicting spectral classifications but our purpose is not to choose the best classification, but only to provide the user with the elements for his own choice.

Approximately 20 % of the cards appear to be redundant, i.e. 30 000 punched cards correspond to approximately 25 000 stars.

1.5. The following step is the addition of information by connection to the data base of the Stellar Data Center (see Ochsenbein et al. 1981, and the communication by Ochsenbein and Bischoff at this Colloquium). We have developed a program which reads each file, goes through the data base, and writes new files containing the following information : identification, equatorial and galactic coordinates, V magnitude, UBV photometry (from Nicolet, 1978), existence of some astrophysical data and reference.

This list can be modified and extended, if necessary.

1.6. Finally it becomes possible to prepare (and to provide to any interested colleague) final editions of the groups on printout, microfiche or magnetic tape. Some of the lists (especially in the later spectral types) have still to be completed, and only partial editions have already been released, namely the catalogue of Be stars (Jaschek and Egret, 1981) and the list of Chemically Peculiar Stars (Egret and Jaschek, 1981), both available on microfiche. The first one includes 1 159 Be stars ; the second one 3 729 He, Ap and Am stars. A general microfiche of the 18 earlier groups (about 6 000 stars) is in preparation. An announcement of the complete catalogue will be published later on, together with definitions of each group and the list of references (Jaschek and Egret, 1982).

SOME QUESTIONS CONCERNING THE C.G.S. :

1.1. Nomenclature :

Although we use preferentially the HD number, 15 % of the stars do not have an HD number. The problem of the identificators is solved through the connection to the data base of the C.D.S.

1.2. Star belonging to more than one group : this situation is acceptable provided that the definitions of the group are compatible and allow such an overlap. For instance, a δ Del star which is a δ Scu variable, or an Ap star which is also a spectroscopic binary. Beside the individual group lists a general index giving for each star the name of the group (s) to which it belongs will also be prepared.

1.3. Completeness : In general we cannot be sure to have an exhaustive sample to a limiting magnitude (e.g. V = 7.0) because of the incompleteness of most of the Surveys. Usually the numbers of objects known differ sharply inside and outside the limits of the Michigan survey by N. Houk, as can be seen for instance in the distribution of the Ap stars given by Egret and Jaschek (1981).

CONCLUSION : USES OF THE C.G.S.

The C.G.S. is of good help for the preparation of observing programs (spectroscopy, photometry, satellite missions, etc...). The availability of extensive samples of peculiar stars makes possible various statistical programs (see for instance the program of absolute magnitude calibration started by Heck et al. 1981). It permits also to eject the peculiar stars when calibrating samples of "normal" stars. The connection with the data base facilitates statistical studies such as the distribution of stars in the Galaxy, the kinematics, the distribution of observational data (photometry, V sini), the location in photometric boxes. In the near future we hope to be able to develope some general algorithms of recognition of peculiar stars by means of a statistical method of data analysis generalizing what is done with the photometric diagrams.

REFERENCES

Egret D. and Jaschek M. : 1981, in "Upper Main Sequence Chemically Peculiar Stars", 23e Colloquium International d'Astrophysique de Liège Univ. de Liège, p. 485

Heck A., Gomez A., Grenier S., Jaschek C. and Jaschek M. : 1981, Bull. Inform. C.D.S. 20, 34

Houk N. and Cowley A.P. : 1975, Michigan Catalogue of Two Dimensional Spectral Types for the HD Stars, vol. 1, Univ. Michigan
Houk N. : 1978, Michigan Catalogue of Two Dimensional Spectral Types for the HD Stars, Vol. 2, Univ. Michigan
van der Hucht K.A., Conti P.S., Lundström I. and Stenholm B. : 1981, Space Science Reviews, 28, 227
Jaschek M. and Egret D. : 1981, in "Be Stars," IAU Symp. 98, M. Jaschek and H.F. Groth ed. p. 261
Jaschek M. and Egret D. : 1982, in preparation
Nicolet B. : 1978, Astron. Astrophys. Suppl. Ser. 34, 1
Ochsenbein F., Bischoff M. and Egret D. : 1981, Astron. Astrophys. Suppl. Ser. 43, 259
Ochsenbein F. and Bischoff M. : this volume, page 211.

CURRENT PROGRESS WITH THE DOUBLE STAR CATALOGUES

Charles E. Worley
U. S. Naval Observatory
Washington. D. C. 20390 U. S. A.

ABSTRACT

Current progress on the visual double star Index and Observation Catalogues is discussed. In addition, a new edition of the visual binary orbit catalogue is being prepared.

A new tape edition of the Index Catalogue is being prepared to replace the previous tape version called the "Index Catalogue, 1976.5". In the last two years, the entire Index Catalogue has been examined and purged of all cases where a star was called double without any supporting evidence. The several thousand such cases are being maintained in a separate file labelled "stars suspected of duplicity". Many other corrections have been made, and the preliminary version of the new tape contains about 73100 objects. After correction of some ordering problems, work will begin on the final version. We hope to be able to provide updated information concerning the motion, interval of observation, and total number of observations for each object. While much of this can be accomplished by computer, human judgement will still have to intervene in a substantial fraction of the work. In the next tape version (1987?), we hope to be able to provide updated magnitudes and spectral types, but this is not a trivial problem.

Since 1973 work has been underway to extend the data in the Observation Catalogue back to the earliest double star data of useable quality, or, approximately, to 1825. This task is virtually completed for the very abundant post-1900 measures, about 80% complete for the 1880-1900 interval, and perhaps 50% finished for the earlier data, leading to an overall completion of about 85%. All major observers have now been included, with the exception of Otto Struve, whose measures require a considerable amount of editorial work. At the moment, approximately 145,000 cards have been punched, and completion of this project should not take

C. Jaschek and W. Heintz (eds.), Automated Data Retrieval in Astronomy, 257–258.
Copyright © 1982 by D. Reidel Publishing Company.

longer than another few years. We will then have a complete, machine-readable data base extending for more than 150 years. Plans are also afoot to transfer both the Index and Observation Catalogues to disk storage, which will permit more rapid access to the data for the purposes of interrogation, correction, and updating.

Eighteen years ago the writer published a catalogue of visual binary orbits, and in 1970 its successor, the Finsen-Worley Catalogue, was published. Computation of new orbits, as well as revisions of old ones, appear to require a new orbit catalogue about every decade; hence a new one is under construction jointly with W.D. Heintz. This catalogue will contain orbits for about 800 visual binaries. Each orbit is being evaluated critically and independently by us, using the extensive data already present in the Observation Catalogue. In nearly every case, we find ourselves in good agreement on the quality rating to be assigned. The catalogue also will contain a brief "for-the-record" list of orbits rejected from the main body of data. These comprise cases where the original computation was seriously erroneous, as well as unconfirmed astrometric pairs. Because of the rapid spread of hand and desk-top computers, we plan to drop the listing of Thiele-Innes elements. Data on MK spectral types and photoelectric magnitudes will be provided by the National Space Science Data Center through the courtesy of W.H. Warren. We anticipate publication of the catalogue within the year.

METACATALOGUE OF GALAXIES AND LARGE SCALE STRUCTURES IN THE UNIVERSE

Kalinkov M.
Department of Astronomy
Bulgarian Academy of Sciences, Sofia

The present state of the compilation of the Metacatalogue of galaxies is described by Kalinkov et al. (1981). The preparation of the catalogue originated from the necessity to put on magnetic tape Zwicky's catalogue in order to study the distribution of galaxies and clusters of galaxies. Enlarging the Metacatalogue by joining to it other catalogues, some interesting results for the large scale structure in the Universe have been obtained. New methods for efficient data processing have been developed for smoothing and filtering of discrete fields, for computing the two-cell correlation function and its matrix representation to test homogeneity and isotropy, and so on (Kalinkov, 1974, 1976, 1977).

It is found that independently of which clusters are used, be it Abell or Zwicky's, the large scale distribution is almost the same. The largest condensations of matter which confidently can be traced have a size of 300 h^{-1} Mpc ($H = 100$ h km s^{-1} Mpc^{-1}) for the Abell and Zwicky catalogues. Moreover the characteristic size of second-order clusters of galaxies is (50 ± 4) h^{-1} Mpc for both Abell and Zwicky clusters, a result which is confirmed by four independent statistical tests (Kalinkov and Kuneva, 1980) and is supported by the analysis of the distribution of Zwicky galaxies. Smoothing and filtering of the Lick counts of galaxies with different functions allows to search aggregates with large density contrast. Most of them have a characteristic size of $50 \div 60$ h^{-1} Mpc.

Adopting a m_{10} - z relation for Abell clusters (Kalinkov et al, 1975) it is possible to compute the two point correlation function according to Peebles (1973, 1980) which is $u = (22.74/R)^{2.10}$. This is the first space correlation function, derived directly without using the angular correlation function.

However, the most interesting result is connected with the hierarchy in the Universe. According to de Vaucouleurs (1970, 1971) there is a density-radius relation for galaxies, (cgs units)

C. Jaschek and W. Heintz (eds.), Automated Data Retrieval in Astronomy, 259–260.
Copyright © 1982 by D. Reidel Publishing Company.

$$\log \rho = 15.19 - 1.7 \log. r$$

and there is no indication for an approach to $\log \rho \simeq$ const out to $\log r \simeq 27$. Data processing of the Metacatalogue shows that this should be the case but for a smaller range. Applying a modification of the nearest neighbour method to the Abell clusters, the relation is

$$\log \rho = 11.98 - 1.617 \log. r.$$

Assuming that a calibration is made for $R = 10\ h^{-1}$ Mpc, requiring that the mean mass of one Abell cluster is $12.43.10^{15}\ M_\odot$ (one order of magnitude higher than the usual estimate) this relation is just the de Vaucouleurs relation, but for $5 < R < 30$. For $R \simeq 100\ h^{-1}$ Mpc which is nearly the limit for the nearest neighbour method, there is a strong approach to $\log \rho \simeq - 29.2$. Our results show that both models of hierarchical clustering of de Vaucouleurs namely that clustering occurs on all scales i) with relative maxima near a series of preferred characteristic sizes (galaxies, clusters, ...) and ii) with no preferred sizes, are not consistent with observations. Another model is more reliable - it is the same as i), but with the addition, that with increasing order of clustering, the density contrast tends to unity.

REFERENCES

De Vaucouleurs, G. : Science, 167, 1203 (1970)

De Vaucouleurs, G. : Publ. Astron. Soc. Pac. 83, 113 (1971)

Kalinkov, M. : Proc. First Europ. Astron. Meet. 3, 142 (Springer Verlag) (1974)

Kalinkov, M. : Mem. Soc. Astron. Ital. 45, 637 (1976)

Kalinkov, M. : Highlights of Astronomy, 4, part I, 279 (1977)

Kalinkov, M., Kuneva, I. : C.r. Acad. Bulg. Sci. 33, 1305 (1980)

Kalinkov, M., Kuneva, I., Stavrev, K. : this Volume (1982), p. 261.

Kalinkov, M., Stavrev, K., Kuneva, I. : Sov. Astron. J. (Lett.) 1, N° 2, 7 (1975)

Peebles, P.J.E. : Astrophys. J. 185, 413 (1973)

Peebles, P.J.E. : The Large-Scale Structure of the Universe (Princeton University Press) (1980)

DATA RETRIEVAL IN THE METACATALOGUE OF GALAXIES

M. Kalinkov - I. Kuneva - K. Stavrev
Department of Astronomy
Bulgarian Academy of Sciences - Sofia

To study the large scale structure of the Universe, we started several years ago the compilation of a new type of catalogue - a metacatalogue of optical extragalactic objects, on magnetic tape. This catalogue may be regarded to some extent as a data bank but it will also include characteristics, reduced to homogeneous systems. The Metacatalogue will contain information for galaxies, clusters of galaxies, counts of galaxies and bibliographical data. Until now some files for individual catalogues and lists have been created, which may be used also before their merging into the Metacatalogue.

1. GALAXIES

Zwicky galaxies (Zwicky et al, 1961-1968). The file contains sequential number, R.A. and Decl. (1950), NGC or IC identification, magnitude and remark, magnitude and classification by Shapley and Ames (1932), Bigay (1951), Pettit (1954), Humason, Mayall and Sandage (1956), Holmberg (1958) (all these catalogues have also been put on magnetic tape), redshift, number(s) of field(s). An abbreviation system is adopted for the remarks for the individual galaxies. E.g. B. = bar, BRS = brightest, CL = cluster, C = compact, DIF = difuse, D = double, DW = dwarf, F = faint, G = galaxy, H = halo, LG = large, MSYS = multiple system, P = pair, R = ring, TSYS = triple system, V = very, and so on.

Zwicky compact galaxies (Zwicky, 1971) - contains sequential number R.A. and Decl. (1950), identification in the seven lists, magnitude.

Uppsala galaxies (Nilson, 1973) - the file includes all data : sequential number, coordinates, identification in other catalogues, PSS number, location in "near" Zwicky clusters, diameters, positional angle, inclination, information parameter, classification by Hubble, van den Bergh and Vaucouleurs, description, magnitude, redshift and notes. The vocabulary of abbreviations for Zwicky galaxies is used, but some other abbreviations are added : ASYM = asymmetric, COND = condensation,

C. Jaschek and W. Heintz (eds.), Automated Data Retrieval in Astronomy, 261–263.
Copyright © 1982 by D. Reidel Publishing Company.

GR = group, SYMB = symbiotic, and so on. In this file an extensive coded system is used to represent a greater amount of information in machine readable form - specially for diameters, classification, magnitudes, and partly for notes.

Vorontsov-Velyaminov galaxies (Vorontsov-Velyaminov et al, 1961-1974) at present contains only sequential numbers by fields, R.A. and Decl. (1950), identification in other catalogues, magnitude, diameters, intensities and inclination (without morphological description and notes).

Many other smaller catalogues and lists of galaxies are being entered on magnetic tape - e.g. Vorontsov-Velyaminov interacting galaxies, Markarian, Arakelian galaxies, ... All data will be merged together with RNGC data, information from de Vaucouleurs and de Vaucouleurs (1964) and de Vaucouleurs, de Vaucouleurs and Corwin (1976) catalogues to compile an extensive part for the galaxies in the Metacatalogue.

2. CLUSTERS OF GALAXIES

Abell clusters (Abell, 1958) contains his number, R.A. and Decl. (1950) m_{10}, distance group, richness group, galactic and supergalactic and Lambert coordinates, corrected cluster radius, mean apparent density, redshift (if any) and distance together with three space components for a newly calibrated m_{10} - z relation.

Zwicky clusters (Zwicky et al, 1961 - 1968) includes sequential number, R.A. and Decl. (1950), type, population, diameter in cm on PSS, distance group, galactic and supergalactic and Lambert coordinates, radius in degrees, mean apparent density and number(s) of field(s).

Some lists for Abell, Zwicky or other clusters of galaxies are being entered on magnetic tape, together with additional information, e.g. Leir and van den Bergh (1977), Dressler (1980).

3.COUNTS OF GALAXIES

An important features of the Metacatalogue is the inclusion of counts of galaxies. We have on magnetic tape the following, starting with the classical counts of Hubble (1934,1936) through Zonn (1968), Rudnicki et al. (1973) to the famous Lick counts (Shane and Wirtanen, 1967) and the reduced Lick counts (Kiang, 1968) and other smaller-area results.

Data retrieval for these three large parts of the Metacatalogue (without references) can be carried out by equatorial, galactic, supergalactic and Lambert coordinates and by identification for any catalogue, included in the Metacatalogue. It is easy to create a subcatalogue by some parameter - magnitude, population, distance group, diameter, mean apparent density and so on. It is also very easy to create counts with any given cells for the galaxies from (1) and to compare with counts from (3).

Some of the files for Abell and Zwicky clusters, cross-identifications, Zwicky galaxies and explanations are in the CDS. We hope to publish in the near future extended explanations for data contents and lists of errors in the Zwicky, Abell, Nilson, Vorontsov-Velyaminov and some other catalogues.

REFERENCES

Abell G.O. : Astrophys. J. Suppl. 3, 211 (1958)
Bigay J.H. : Ann. Astroph. 14, 319 (1951)
de Vaucouleurs, G., de Vaucouleurs A. : Reference Catalogue of Bright Galaxies, Univ. of Texas Press, Austin (1964)
de Vaucouleurs G., de Vaucouleurs A., Corwin H.G. : Second Reference Catalogue of Bright Galaxies, Univ. of Texas Press, Austin (1976)
Dressler A. : Astrophys. J. Suppl. 42, 565 (1980)
Holmberg E. : Medd. Lunds Astr. Obs. Ser. II, Nr 136 (1958)
Hubble E. : Astrophys. J. 79, 8 (1934)
Hubble E. : Astrophys. J. 84, 517 (1936)
Humason M.L., Mayall N.U., Sandage A.R. : Astron. J. 61, 97 (1956)
Kiang T. : Dunsink Obs. Publ. 1, 109 (1968)
Leir A.A., van den Bergh S. : Astrophys. J. Suppl. 34, 381 (1977)
Nilson P. : Uppsala General Catalogue of Galaxies, Uppsala (1973)
Pettit E. : Astrophys. J. 120, 413 (1954)
Rudnicki K. et al : Acta Cosmologica 1,7 (1973)
Shane C.D., Wirtanen C.A. : Publ. Lick Obs. 22, 1 (1967)
Shapley H., Ames A. : Ann. Harv. Coll. Obs. 38, N° 2 (1932)
Vorontsov-Velyaminov B. et al. Morphological Catalogue of Galaxies, I-V, Moscow University (1961-74)
Zonn W. : Acta Astron. 18, 273 (1968)
Zwicky F. : Catalogue of Selected Compact Galaxies and of Post-Eruptive Galaxies, Speich, Zürich (1971)
Zwicky F. et al. : Catalogue of Galaxies and of Clusters of Galaxies I-VI, CIT, Pasadena (1961-68)

LE CENTRE DE DONNEES DES AMAS DE GALAXIES

A. Fernandez[+], G. Mathez[+], A. Mazure[+], L. Nottale[+], R. Rothenflug[‡]
+ Observatoire de Meudon ‡ C.E.N. Saclay

Abstract : A Cluster of Galaxies Data Center is in the process of being settled. Its aim will be to collect the data presently scattered in the literature, concerning clusters of galaxies, in all wavelength ranges. We describe its diverse entries and we discuss some specific problems and its possible future utilization.

1. INTRODUCTION

Alors qu'un certain nombre de catalogues de base d'amas de galaxies ont été établis (Abell, 1958, Zwicky et al., 1961-1968 ; amas du Sud : Klemola, 1969, Snow 1970, Sersic, 1974, Rose, 1976, Duus et Newell, 1977, Braid et Mac Gillivray, 1978 ...), il n'existe pas jusqu'à présent de catalogue général qui contienne tous les amas répertoriés ainsi que les données, actuellement éparses dans la littérature, les concernant. Pour certaines données, un tel travail de compilation a déjà été fait, par exemple les décalages spectraux moyens (Noonan, 1981), mais il demande de toutes façons une remise à jour périodique. Le but du Centre de Données des Amas de Galaxies (CEDAG) est de stocker les données, non traitées dans un premier temps, et leurs références, existant pour les amas dans tous les domaines de longueur d'onde. Il comprendra aussi bien les superamas que les groupes et les amas proprement dits. Deux types de données sont à distinguer : les grandeurs caractéristiques de l'amas qui sont données par les catalogues de base et qu'on peut en première approximation considérer comme fixées (exemple : classes de richesse, de compacité) et les grandeurs qui dépendent plus de l'évolution des observations (distribution des décalages spectraux, détection d'un milieu intergalactique ...).

2. DESCRIPTION DU CATALOGUE

Les catalogues existant en formeront l'ossature. Un amas sera identifié par les coordonnées équatoriales α, δ de son centre en h.mn x d°arcmin.

C. Jaschek and W. Heintz (eds.), Automated Data Retrieval in Astronomy, 265–267.
Copyright © 1982 by D. Reidel Publishing Company.

2.1. Désignations diverses

Les noms de l'amas dans tous les catalogues de base où il est répertorié apparaitront dans cette rubrique (exemple : numéro d'Abell, numéros de Zwicky) ainsi que ses coordonnées. Les références de cross-identification entre catalogues seront données. Les noms du superamas auquel un amas appartient, ou des amas ou groupes qu'il contient éventuellement seront stockés, ce qui comprend les cas où plusieurs amas d'Abell sont inclus dans un amas de Zwicky.

2.2. Grandeurs caractéristiques. Données de catalogues

Cette rubrique comprendra les grandeurs caractéristiques de l'amas, données en général dans les catalogues sous forme de classes : le diamètre angulaire (Zwicky, autre catalogue) ; un décalage spectral moyen indicatif (3 décimales) ; la classe de distance (Abell, Zwicky, autre catalogue) ; la classe de richesse (classe d'Abell, population de Zwicky, autre) ; des magnitudes représentatives (galaxie la plus brillante, 10ᵉ galaxie d'Abell, autre magnitude caractéristique) ; des classifications morphologiques (compacité de Zwicky, classes de Bautz-Morgan, de Rood-Sastry).

2.3. Objets particuliers

Les noms et/ou les coordonnées d'objets particuliers appartenant à un amas seront stockés : galaxies d'Arakelyan, galaxies parentes de supernovae, galaxies compactes, galaxies de Markarian, galaxies de Seyfert, galaxies cD, galaxies X, "head-tail" galaxies, objets de Vorontsov-Velyaminov, objets de Arp, galaxies N, quasars.

2.4. Détection de halo

Cette rubrique indiquera brièvement si le halo d'un amas a été recherché et éventuellement détecté, dans tous les domaines de longueur d'onde, radio, IR, optique, X et γ, en donnant les références correspondantes.

2.5. Distribution des vitesses radiales

La vitesse radiale moyenne, la dispersion estimée des vitesses ainsi que le nombre de galaxies de l'amas de vitesses connues seront stockés. Des références éventuelles analysant plus en détail la distribution des vitesses dans l'amas seront répertoriées.

2.6. Données optiques

Cette rubrique comprendra les références sur la morphologie des galaxies de l'amas, la répartition des types morphologiques, les références sur la photométrie des galaxies de l'amas, sur sa fonction de luminosité, sur son profil de densité et de luminosité ; on indiquera la luminosité d'un milieu intergalactique éventuel, ainsi que les caractéristiques morphologiques globales de l'amas.

2.7. Données X

Les données stockées concerneront les grandeurs suivantes : boite d'erreur dans une bande d'énergie donnée ; distance centre optique-boite d'erreur ; flux et température dans une bande d'énergie donnée ; présence éventuelle d'une raie et l'énergie qu'elle contient ; caractéristiques morphologiques (dimension, structure, cartes ...).

2.8. Données radio

On indiquera, à 21cm et aux autres longueurs d'onde étudiées, la présence éventuelle d'un fond diffus continu ou dans la raie 21cm, et les noms et/ou coordonnées de radiogalaxies appartenant à un amas.

3. DISCUSSION

La constitution du catalogue est en cours. Elle appelle plusieurs remarques. Tout d'abord un certain nombre de problèmes spécifiques liés à la nature même des amas de galaxies sont rencontrés : ce sont des objets composites dont certaines caractéristiques peuvent dépendre fortement des critères utilisés pour les définir. Ceci implique le choix de stocker les données non traitées ainsi que les conditions précises de leur obtention. Dans le futur, le catalogue sera informatisé et une homogénéïsation statistique de certaines données pourra être effectuée. Quelques exemples d'utilisation possible du catalogue peuvent être indiqués : corrélations entre différents domaines de longueur d'onde (exemple luminosité X - densité centrale ou pourcentage de spirales, température X - dispersion des vitesses) ; cosmologie (évolution de grandeurs caractéristiques en fonction du décalage spectral) ; aide aux observations ...

Références :

Abell, G.O., 1958, Astrophys. J. Suppl. 3, p.211
Braid, M.K., Mac Gillivray, H.T., 1978, Mon.Not.R.astr.Soc.182, p.241
Duus, A., Newell, B., 1977, Astrophys. J.Suppl.35, p.209
Klemola, A.R., 1969, Astron.J.74, p.804
Noonan, T.W., 1981, preprint
Rose, J.A., 1976, Astron.Astrophys.Suppl.23, p.109
Sersic, J.L., 1974, Astrophys.Space Sci.28, p.365
Snow, T.P.Jr, 1970, Astron.J.75, p.237
Zwicky, F., Karpowicz, M., Kowal, C.T., Herzog, E., Wild, P., 1961-1968, "Catalogue of Galaxies and of Clusters of Galaxies", California Institute of Technology

CENTRALIZATION OF EXTRAGALACTIC DATA

Paturel G.
Observatoire de Lyon

Abstract - We present a project of centralization of extragalactic data concerning normal galaxies.

The project of creating a data bank about galaxies is an old dream. Knut Lundmark had already thought of such a project, but the work never came to anything. The use of computers has made this work possible.

We shall try to answer four questions :

I) Why centralize data ?

Three reasons to support this project :

1) Any statistical study begins with the tedious work of constructing sample. It is better to record this information at the start so that it is available to everyone.

2) The centralization makes it possible to compare the different sources of data. It is then possible to judge the samples better and to improve them by "homogeneization" methods (ex. : reduction of isophote diameters to a given system).

3) New statistical methods for data analysis permet studies which were impossible before (Taxonomy or principal component analysis). These studies require large samples containing numerous parameters.

II) How to do such a centralization ?

It could be difficult to create a specialized center with enough staff to permet at the same time the collecting of new data, its treatment for homogeneization and diffusion.

It now seems easier to group from time to time the compilations made by different astronomers. We already have working collaboration between different groups.

This could lead to the periodical publication of an up to date catalog containing parameters as homogeneous as possible.

At the moment only the centralization of numerical data and associated references is possible (for example the collection of isophote charts can not be considered).

C. Jaschek and W. Heintz (eds.), Automated Data Retrieval in Astronomy, 269–271.
Copyright © 1982 by D. Reidel Publishing Company.

TABLE 1

Principal new general catalogs and compilations about normal galaxies.

Short designation	Reference	Nbr of galaxies	criteria	parameters
UGC	Nilson (1973)	12921	$\delta > -2°30'$ $D > 1'$	α, δ, D, D, T, m (V), β
EUS	Holmberg et al. (1981)	18500	$\delta < -18°$ $D > 1'$	α, δ, D, T, β
CGCG	Zwicky et al. (1968)	31350	$\delta > -3°$ $m < 15.7$	α, δ, m, (V)
MCG	Vorontsov-Veljaminov (1974)	34000	$\delta > -45°$	α, δ, T, m
SGC	Corwin (198?)		$\delta < -18°$ later $\delta < +3°$ $D > 1.5'$	α, δ, D, T ...
RC2 (*)	Vaucouleurs et al. (1976)	4364	No	(α, δ, T, L, D, D, m, C, SH, V)
HICG (*)	Bottinelli et al. (1981)	1200	No	W, SH, V
IGS (*)	Gisler et al. (1979)	2004	$\delta > 3°$	V
(*)	Rood (198?)		No	V

(*) Compilation

α, δ, coordinates
β, position angle
D, major and minor axes
T, morphological type
L, luminosity class

m, magnitude
V, radial velocity
C, colors
SH, neutral hydrogen flux
W, 21-cm line width

III) Where to find available data ?

1) The general catalogs group all known objects relating to certain criteria (coordinates, magnitudes, diameters ...). We give in table 1 a list of principal general catalogs about normal galaxies.

2) Compilations. These are the collection of all existing data for the largest possible number of galaxies. We give in table 1 the principal compilations.

3) Papers about new measurements. There are a large number of such lists. The information can concern one or many galaxies.

IV) What can be the role of the IAU ?

IAU can play an important role by publishing some recommendations about the following points :

1) Nomenclature. (ex. : NGC, IC and Anonymus; the anonymus number being followed by nine digits).

2) Publication of raw data without corrections.

3) Publication of astronomical and technical characteristics such as the coordinates as accuratly as possible, the name of the telescope, the velocity resolution and spatial resolution ...)

4) Use of units and notations as homogeneous as possible.

Even within the proposed solution it is necessary to get technical assistance and IAU can support the application to directors for financial assistance to continue the project.

References

Bottinelli, L., Gouguenheim, L., Paturel, G., (1981), Astron. Astrophys. Sup. (in press)

Gisler, G.R., Friel, E.D., (1979), Index of Galaxy Spectra, ed. Pachart, Tucson

Holmberg, E.B., Lauberts, A., Schuster, H.E., West, R.M., (1974-1981), Astron. Astrophys. Sup. 18, 463, 18, 491, 22, 327, 22, 295, 31, 15, 34, 285, 39, 173, (in press), (in press)

Nilson, P., (1973), Uppsala General Catalog of Galaxies, Uppsala Astr. Obs. Annales, 6

Vaucouleurs, G. de, Vaucouleurs, A. de, Corwin, H., (1976), Second Reference Catalog of bright Galaxies, Univ. of Texas Press Austin

Vorontsov-Veljaminov, B.A., Arkipova, A.A., Krasnogorskaja, A.A., (1963-1974), Morphological Catalog of galaxies, Trudy Sternberg State 33, 34, 38, 32, 46

Zwicky, F., Herzog, E., Karpowicz, M., Kowal, C.T., (1962-1968), Catalog of galaxies and clusters of galaxies, California Institute of Technology, Pasadena.

THE COMPILED CATALOGUE OF GALAXIES IN MACHINE-READABLE FORM AND ITS STATISTICAL INVESTIGATION

N.G. Kogoshvili
Abastumani Astrophysical Observatory
Georgia - U.S.S.R.

The compilation of a machine-readable catalogue of relatively bright galaxies was undertaken in Abastumani Astrophysical Observatory (Kogoshvili 1972 ; 1975) in order to facilitate the statistical analysis of a large observational material on galaxies from the Palomar Sky Survey. This catalogue is mainly based on the data from the MCG catalogue (Vorontsov-Veljaminov et al, 1962-1968) supplied with information from the UGC (Nilson, 1973), RCBG (De Vaucouleurs and De Vaucouleurs, 1964), CGCG catalogues (Zwicky et al, 1961-1968) and from other sources.
In compiling the catalogue of galaxies the following problems were considered :

1. the collection of existing information for each galaxy ;

2. a critical approach to data aimed at the selection of the most important features of the galaxies ;

3. the recording of data in computer-readable form, and

4. the permanent updating of the catalogue.

Thus the Abastumani Catalogue comprises about thirty three thousand galaxies most of them brighter than $15.^{m}1$ and located in the area north of -33°.

The catalogue provides for each galaxy : MCG and NGC numbers, equatorial coordinates, integrated magnitude, diameters and surface brightnesses for the internal and external areas of the galaxy, inclination and detailed description of the galaxy by symbols. All these data are given according to MCG. Also for some galaxies the following data are given : different morphological classifications of galaxy types, integrated colour indices, radial velocities, the occurrence of a galaxy in some lists of galaxies, the membership of a galaxy in binary and multiple systems and other parameters.

C. Jaschek and W. Heintz (eds.), Automated Data Retrieval in Astronomy, 273–275.
Copyright © 1982 by D. Reidel Publishing Company.

A broad range of statistical investigations was undertaken by means of the Compiled Catalogue of Galaxies revealing some systematic data errors of the catalogues used.

In particular, as a result of the statistical investigations of the dependence between the visibility of a galaxy and its position on the Palomar Sky Atlas maps, on the basis of MCG and UGC data, the "boundary field effect" was studied. The analysis of counts of galaxies in the MCG catalogue showed a deterioration of the visibility of galaxies all over the eastern half of the field and particularly in the east boundary region as compared with the western region. This effect is most appreciable in the distribution of galaxies with brightness fainter than the limiting magnitude of MCG, as well as in the distribution of elliptical galaxies. Most likely this effect comes from a weak asymmetry of the photographic field of the 48" Schmidt telescope.

Data from the UGC catalogue, which similarly to the MCG catalogue is based on the Palomar Sky maps, confirm the reality of the field effect studied. This supports the view that the boundary effect is possibly due to the optical system of the Palomar Schmidt telescope.

A computer made analysis of the irregularities in the distribution of the numbers of galaxies in neighbouring fields made it possible to estimate the value of the relative mean plate-to-plate fluctuation of the number of galaxies for the MCG data. This fluctuation corresponds approximately to the contribution of observing condition variability into the values of the two-point correlation function of the galaxy numbers obtained by Peebles.

Data of the Compiled Catalogue made it possible to study some features of Zwicky's compact galaxies and compare them with those of galaxies from MCG and UGC, selected by the compactness criterion.

An investigation of the surface brightness of Zwicky's compact galaxies showed that only 2 % have the surface brightness brighter than $20.^{m}0/''$ and it is hardly possible to regard the considerable number of Zwicky's objects as compacts based on the compactness criterion suggested by Zwicky.

The study of the mean value of the surface brightness for the galaxies of MCG showed that about thousand galaxies in the catalogue have small dimensions, high surface brightness (brighter than $21.^{m}0/''$) and spherical appearance, suggesting thus that these objects could belong to the class of compact galaxies.

We also investigated the problem of the small-scale clustering of compact galaxies on the basis of the Zwicky and MCG lists. With this aim the frequency distribution of the apparent angular separations between compact galaxies and their nearest neighbours as projected on the celestial sphere was studied by means of a statistical technique.

The analysis of the counts showed that the clustering tendency among Zwicky's galaxies is connected with eruptive activity rather than with surface brightness. With regard to the galaxies with high surface brightness in the MCG catalogue, it is hardly possible to assume now that compact galaxies occur more frequently in systems of galaxies than normal galaxies, due to systematical errors in estimating the angular dimensions and magnitudes for fainter objects.

Using the MCG data, the problem of small-scale clustering of the spiral and elliptical galaxies was studied for different intervals of magnitude, as well as the problem of the coiling of spiral components of apparent pairs based on a larger amount of data then in previous publications.

Thus we summarize the conclusions as follows :

1. A catalogue of galaxies was compiled on magnetic tape, which is to be permanently updated. This catalogue may be useful as a basis for solving a wide range of statistical extragalactic problems.

2. On the basis of the data of the Compiled Catalogue of Galaxies the value of the irregularity in galaxies distribution within a single plate of the PSA, as well as the plate-to-plate irregularity was statistically estimated. The significant distortions of the apparent distribution of galaxies up to 15^m should be considered in constructing models of space distribution of galaxies.

3. Selection of compact galaxies by the surface brightness criterion leads to systematic overestimates of the number of compact galaxies as their brightness decreases, and to increase therefore their tendency of clustering.

4. The analysis of the vast data of MCG confirms the opinion, that E-galaxies occur more frequently in pairs as compared with spirals, and also the opinion concerning a larger frequency of pairs with opposite directions of coiling as compared with pairs whose components have the same spiral arm coiling direction.

REFERENCES

Kogoshvili N.G. : Astron. Circ. Ac. Sci. USRR 706, pp 1-3 (1972)
Kogoshvili N.G. ; Abastumansk. Obs. Bull. 46, pp. 133-162 (1975)
Nilson P. : Uppsala General Catalogue of Galaxies Uppsala (1973)
Vaucouleurs G. (de) and Vaucouleurs A. (de) : Reference Catalogue of Bright Galaxies - Univers. Texas (1964)
Vorontsov-Veljaminov B.A. , Krasnogorskaya A.A. and Arkhipova V.P. : Morphological Catalogue of Galaxies - Moscow - (1962-1968)
Zwicky F., Herzog E., Wild P., Karpowicz M. and Kowal C. : Catalogue of Galxies and of Clusters of Galaxies - Zurich (1961-1968)

GENEVA PHOTOMETRIC BOXES. IV. A REFINED METHOD FOR DIRECT ACCESS

Bernard Nicolet
Observatoire de Genève
CH-1290 Sauverny, Switzerland

Abstract. In the Geneva photometry, a photometric box around a so-called central star is the set of stars having almost the same colours as this central star.

For the most current applications, it is useful to construct the boxes of a few hundred stars. A method allowing to obtain these boxes rapidly is described. It is based on so-called direct access square files. A FORTRAN routine on a UNIVAC 1160 computer is widely used at the Geneva Observatory. The principles described in this paper may be easily adapted to build square files and to write routines in evolved languages for most computers.

Other applications of square files are also outlined.

1. INTRODUCTION

The Geneva photometric boxes were initially described by Golay et al. (1969, 1977). Nicolet (1981a) has extended this description. Their homogeneity as tested by Golay (1978a), Crézé et al. (1980) and Nicolet (1981a) allows various applications such as calibration of open clusters (Golay 1973, 1978b, Nicolet 1981c), estimations of faint reddening (Nicolet 1981b), reddening laws in the UV (Nicolet 1982) and so on.

Golay and Mandwewala (1977, 1978) have prepared two catalogues of photometric boxes based on the second Geneva photometric catalogue by Rufener (1976). A new catalogue (Nicolet 1981d) is almost ready from the third Rufener (1981) catalogue. It will be available on magnetic tape and microfiches (no printed version) and announced in the Supplements of Astronomy and Astrophysics.

The aim of this paper is the description of a method allowing to obtain fairly fast boxes for some central stars.

C. Jaschek and W. Heintz (eds.), Automated Data Retrieval in Astronomy, 277–280.
Copyright © 1982 by D. Reidel Publishing Company.

2. PRELIMINARY CONSIDERATIONS

The principles used to construct photometric boxes are quite simple:

a) Sort the Rufener (1981) catalogue using the parameter Δ as a key. Recall that $\Delta=(U-B_2)-0.832\ (B_2-G)$ is almost insensitive to reddening.

b) Find the central star with its Δ.

c) Select in the sorted file candidates having a Δ compatible with belonging to the box.

d) Examine each candidate with all criteria.

A question remains: how to organize the sorted file?

a) Direct access. Advantage: the only candidates (typically 200) are read. Drawback: Each direct access reading needs 0.1 sec on a UNIVAC 1160, hence the construction of a box needs about 20 sec of I/O time.

b) Sequential access. The software of most computers (e.g. our UNIVAC 1160) allows a very fast sequential reading of a file: typically 0.002 sec per record. On average half of the file (~7500 records) is read. The CPU time is also non-negligible. The construction of a box needs up to 30 sec.

3. SQUARE DIRECT ACCESS FILES

It is possible to combine the advantages of the sequential and direct access. In principle each record includes the information for many stars and each reading provides many candidates, hence the I/O time is divided by the number of stars in a record.

In our method, we use square direct access files. The number of words (36 bits for the UNIVAC 1160) is equal to the number of records whence the adjective "square". The last record plays the role of contents and its ith word is the key-parameter Δ of the first star in the ith record. This last record is read once and kept in the memory.

As a numerical example, let us describe the square file used to construct the Geneva photometric boxes.

Number of stars:	15000 (Rufener 1981)
Number of words per star (packed form):	11 wds
Length of a record:	440 wds
Number of records:	440
Number of candidates per reading:	40 = 440/11

Hence 2-4 readings per box
880 words in the memory (last record + current record)
0.3 to 0.5 seconds computer time per box.

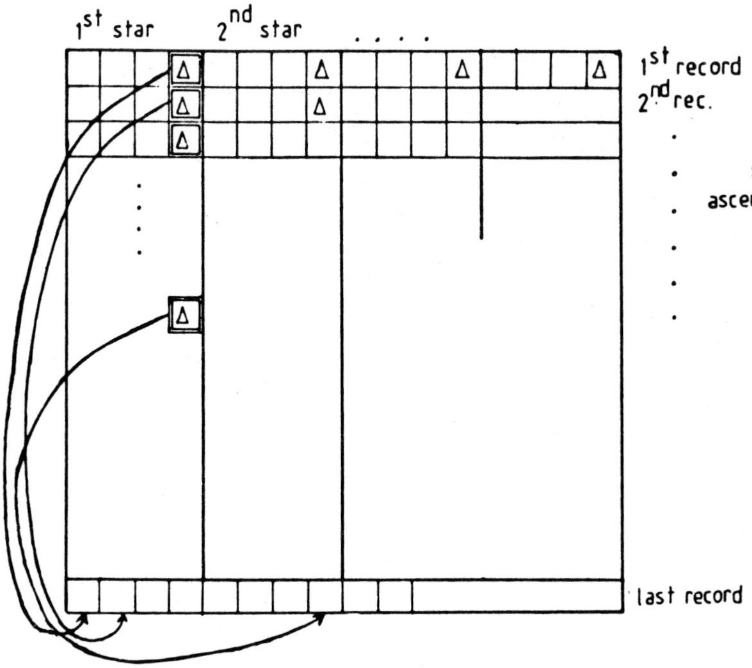

Having the central star, we may obtain its colours and quickly select our candidates

colours

Δ : key parameter

last record

n=n(Δ−ε) ε : size of the box

read record n

j = 0

j = j+1

n=n+1 Yes $j \overset{?}{>} 40$

No No

$\Delta_{cand} \overset{?}{>} \Delta+\varepsilon$ No Other criteria? Yes Box's star

Yes

End of box

4. OTHER APPLICATIONS

The same procedure may of course be applied to other photometric systems. But "coordinate boxes" may also be useful

a) to solve some identification problem

b) to draw identification charts

It is advisable to avoid the right ascension α as a key parameter (polar regions!) δ or more refined key (such as the celestial cube) are more advisable.

REFERENCES

Crézé, K., Turon-Lacarrieu, C., Golay, M., Mandwewala, N.: 1980, Astron. Astrophys. 85, 311
Golay, M.: 1973, IAU Symp. No. 54, p. 27
Golay, M.: 1978a, Astron. Astrophys. 62, 189
Golay, M.: 1978b, IAU Symp. No. 80, p. 277
Golay, M., Mandwewala, N.: 1977, Publ. Obs. Genève Série B, No 4
Golay, M., Mandwewala, N., Bartholdi, P.: 1977, Astron. Astrophys. 60, 181
Golay, M., Peytremann, E., Maeder, A.: 1969, Publ. Obs. Genève, Série A No 76, 44
Nicolet B.: 1981a, Astron. Astrophys. 97, 85
Nicolet B.: 1981b, to appear in Astron. Astrophys. Suppl.
Nicolet B.: 1981c, to appear in Astron. Astrophys.
Nicolet B.: 1981d, to be announced in Astron. Astrophys. Suppl.
Nicolet B.: 1982 , in preparation
Rufener F.: 1976, Astron. Astrophys. Suppl. 26, 275
Rufener F.: 1981, to appear in Astron. Astrophys. Suppl.

DATA FOR THE COMPILATION OF THE " THIRD CATALOGUE OF NEARBY STARS"

Wilhelm Gliese
Astronomisches Rechen-Institut
Heidelberg, Federal Republic of Germany

In 1957 I have published a catalogue of 915 stars - single and multiple systems - nearer or apparently nearer than 20 parsecs. The 1969 edition was extended to 22.5 parsecs with 1328 objects. In the last decade new trigonometric parallax programs have yielded first results and their accuracy has been increased. In 1981 the number of stars with known parallaxes exceeding 0".044 is estimated close to 1700. Certainly, that are not yet thirty per cent of the stars really existing in this volume. But time seems to be ripe for a third edition which should be extended to 25 parsecs now.

Probably, this new catalogue would close a period of parallax determinations. In the near future we expect a remarkable increase in accuracy of trigonometric parallaxes as Harrington and Westerhout say (1981) "Nearby space may no longer include the region within 25 parsecs, but that within 250 parsecs ....".

The goal of our catalogue is on one hand the compilation of known data for investigations of various problems, on the other hand it is a documentation of still insufficient knowledge to stimulate further observations.

The columns give the characteristic data of the objects: position, motion, parallax, spectral type, photometry, and by all means an appendix with Notes to the members of the star list.

The size of the parallax of a star decides for or against its inclusion in the catalogue. The new General Catalogue of Trigonometric Parallaxes (GCTSP) in progress at Yale shall be the basis of the 3rd Catalogue of Nearby Stars. Certainly, also among the stars of this catalogue we shall not find a representative sample of the objects in the solar neighbourhood. Many trigonometric programs have preferred the stars with large proper motions. Such selection effects are reduced by inclusion of nearby objects with spectroscopic and/or photometric parallaxes which have been determined without knowledge of proper motions.

C. Jaschek and W. Heintz (eds.), Automated Data Retrieval in Astronomy, 281–283.
Copyright © 1982 by D. Reidel Publishing Company.

It is indispensable to give spectral types and photometry in the 3rd catalogue of nearby stars and it is desirable to give these quantities in a uniform system - if possible. MK types are available for the brighter stars - we shall take advantage of the compilations at the CDS. For fainter HD stars in southern declinations the catalogues by N. Houk will be used. But we have to consider possible differences between the MK type-luminosity relation based on objective prism classifications and those based on slit spectra. For the largest group in the solar neighbourhood, the faint red dwarf stars, two significant classification systems shall be used also for the 3rd edition: 1) Mount Wilson types found in the Radial Velocity Catalogue (Wilson, 1953) and in the well-known list by Joy and Abt (1974), and 2) the spectral types by Kuiper which are still scattered in the literature.

A few smaller lists of spectral types can be used for luminosity estimates, at least to detect suspected nearby objects. The utilization of all possible spectral type-luminosity relations for distance determinations means that besides all help by data centers we have to search the modern literature. Today we have various photometric systems whose magnitudes and colours make it possible to derive photometric parallaxes. In the columns of the new catalogue we shall give broad-band UBV,RI. These data are the only parameters which are fairly complete and homogeneous for the catalogue stars. Most of these data we shall receive from the CDS, mainly collected by the colleagues at Geneve. Distances based on other systems as Geneve, Strömgren will be taken into consideration too. For degenerates we intend to incorporate also Greenstein's photometric parallaxes.

For many objects various distance determinations will be available but users of the catalogue would like to find the most reliable parallax value for each star: one "resulting parallax" on which are based the luminosity M_V and the space velocity components of this object. It is one of the great difficulties in the compilation of such a catalogue to find the most reliable value out of various series of observations - should we use computed weighted means or should we give priorities ?

Furthermore, our work would be facilitated by a compilation of all the spectral type-luminosity relations and colour-luminosity relations scattered in the literature which can be applied for distance determinations of nearby stars. The various spectroscopic and photometric M_V are certainly not independent from each other. Their apparently best value has to be combined with the trigonometric M_V - this procedure is the other great difficulty when preparing the nearby star catalogue.

Besides these real problems the compilation of the positions and the proper motions makes no serious difficulties, except for a few faint objects. But unfortunately no complete up-to-date catalogue of the radial velocities is available.

A catalogue of nearby stars will be incomplete if it is not supplemented by detailed "Notes" informing on duplicity, on multiple systems, on common proper motion pairs, on variability, on other peculiarities, or even on very discordant observational data for the various objects. This compilation will be facilitated by the data centers. But also continuous examination of current publications is necessary, a work which gets beyond one person. The third edition of the catalogue of nearby stars will be compiled by Gliese and Jahreiss; M. Petit will contribute the data of observed variability.

I close this report with thanks to all colleagues who have supported our work and who will do it in future.

REFERENCES

Harrington,R.S., and Westerhout,G.: 1981, Bull. Inform. CDS 20, pp. 24-27.
Joy,A.H., and Abt,H.A.: 1974, Astrophys. J. Suppl. 28, pp. 1-18.
Wilson,R.E.: 1953, General Catalogue of Stellar Radial Velocities, Carnegie Inst. Washington Publ. 601.

FUNDAMENTAL DATA FOR FK4/FK4SUP STARS

H. Jahreiss
Astronomisches Rechen-Institut, Heidelberg
Federal Republic of Germany

1. INTRODUCTION

The stars in the FK4- and FK4Sup-Catalogue are supplied with visual magnitudes and spectral types essentially on the basis of the Henry Draper Catalogue. There are several good arguments, to compile additional astrometric, photometric and spectroscopic data for these stars. Not only to detect astrometric effects, which can produce errors in the fundamental system, like magnitude equation, colour equation, and foreshortening effect. For the last one distances and radial velocities have to be known. But also to study the kinematics of the FK4/FK4Sup stars, in profiting from their high precision in proper motion.

The hitherto existing compilations of astrophysical data restricted themselves on subsets of the FK4/FK4Sup stars or are out of date. Therefore a new compilation was started. The search for data was much facilitated by the CDS supplying us with the desired catalogues. Throughout this paper it is evident, that in the various observing-programs preference was given to the brighter stars. Since the 1987 FK4Sup stars are concentrated around $m_V = 6^m$, whereas half of the 1535 FK4 stars are brighter than $m_V = 5^m$, the astrophysical data were found to be more complete for FK4 stars than for FK4Sup stars.

2. TRIGONOMETRIC PARALLAXES

Modern observing programs for the determination of trigonometric parallaxes concentrate mainly on faint stars. Therefore the General Catalogue of Trigonometric Stellar Parallaxes and its Supplement (Jenkins 1952, 1963) remain still the main sources. Since 1963 parallaxes were measured for only 92 FK4 stars and 19 FK4Sup stars respectively. Up to date for altogether 1011 FK4 stars and 531 FK4Sup stars published parallaxes could be found. But, due to the large mean errors, we can state, that only 50% FK4 stars and 16% FK4Sup stars have usable trigonometric parallaxes. This situation cannot be improved unless HIPPARCOS will be launched in the mid eighties. So, in determining

C. Jaschek and W. Heintz (eds.), Automated Data Retrieval in Astronomy, 285–287.
Copyright © 1982 by D. Reidel Publishing Company.

distances for the FK4/FK4Sup stars, we are largely dependent on photometric and spectroscopic data.

3. PHOTOMETRY

In UBV-photometry relative recent compilations are available by means of the Geneva catalogues (Nicolet 1975, 1978; Mermilliod 1980). The same is valid for the Geneva-Photometry (Rufener 1980) and for the uvbyβ-Photometry (Hauck and Mermilliod 1980). The latter one was used by Davis Philip and Egret to estimate several astrophysical parameters (M_V, Θ, log g, [Fe/H]). In addition Neckel, Klare and Sarcander published extinction data for early type stars. Altogether, the following photometric data are available,

UBV-Photometry	for	1485	FK4(97%)	and	1605	FK4Sup(81%)
Geneva-Photometry for early type stars	for	1284	FK4(84%)	and	1188	FK4Sup(60%)
uvbyβ-Photometry	for	778	FK4(93%)	and	804	FK4Sup(80%)
Estimated Parameters	for	685	FK4(82%)	and	718	FK4Sup(71%)
Extinction Data	for	623	FK4(74%)	and	406	FK4Sup(40%) stars.

Hereby, a lot of stars lacking UBV-photometry were measured in one or another photometric system.

4. MK SPECTRAL TYPES

The main sources for MK spectral types are the Catalogue of Selected Spectral Types in the MK System (Jaschek 1978), and the first two Volumes of the Michigan Catalogue of Two-Dimensional Spectral Types for the HD-stars (Houk and Cowley 1975, Houk 1978) containing all HD-stars south of $\delta = -40°$. As result there are in the

Cat. of Selected MK-Types	1329	FK4	and	1037	FK4Sup stars,
Michigan Cat. only ($\delta \leq -40°$)	31	FK4	and	304	FK4Sup stars,
other sources (mainly after 1976)	86	FK4	and	123	FK4Sup stars.

Therefore, 94% of the FK4 stars and 74% of the FK4Sup stars are supplied with MK-types. Even selected they are of very different quality.

5. RADIAL VELOCITIES

Two bibliographies comprise the determinations of radial velocities before 1977, Abt and Biggs (1972), and Barbier (1977). Up to now for 1464 FK4 stars (95%) and 1504 FK4Sup stars (76%) radial velocities are published, and a great deal of the remaining FK4/FK4Sup stars were put on running observing-programs. But, comparable to the MK-types, the data are of different accuracy, and to date a necessary renewal of Wilson's system in the General Catalogue of Stellar Radial Velocities

(1953) is not in sight.

6. CONCLUSION

The present results display an increasing amount of photometric and spectroscopic data for the brighter stars. 1461 FK4 stars (95%) and 1616 FK4Sup stars (81%) are in Hoffleit's Catalogue of Bright Stars. But, whilst for the trigonometric parallaxes a new system will be available in the near future, the lack of homogeneous spectroscopic data will continue.

REFERENCES

All the Catalogues quoted are in the list of available Catalogues at the CDS. For further references see
Jahreiss,H.: 1981, Bull. Inform. CDS 10, 13

REMARKS ABOUT THE CATALOGUING OF OPEN CLUSTER DATA

Gösta Lyngå
Lund Observatory
Box 1107
S-221 04 Lund
Sweden

During the last three years I have collected and examined data for open clusters. I have chosen to make a computer-based catalogue. Fig. 1 shows the structure of this work. The data reside on a disc at the Lund Observatory. As original input I have used the data banks available in 1980. Later information is added through frequent updates. There are two modes of retrieving the data: by listing and by direct access.

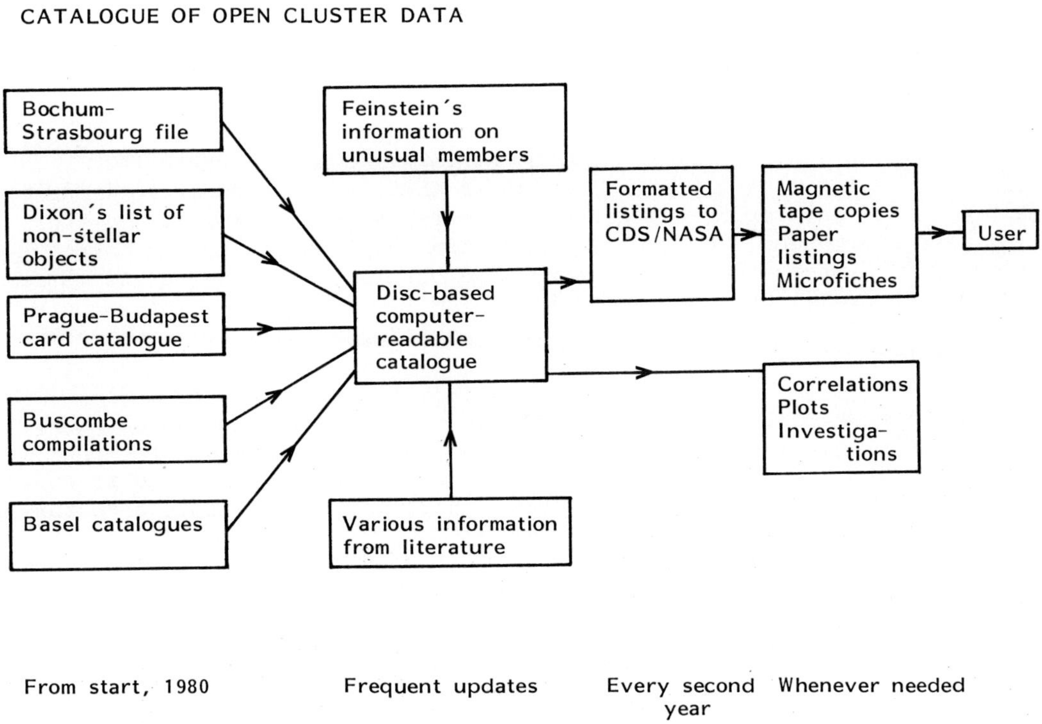

Fig. 1. Cluster catalogue lay-out.

C. Jaschek and W. Heintz (eds.), Automated Data Retrieval in Astronomy, 289–291.
Copyright © 1982 by D. Reidel Publishing Company.

I. LISTING OPTIONS

One way of retrieving the data is to make a listing and I do so every second year in order to distribute the data as it is known by the computer base at that stage. The distribution goes via CDS and NASA and can be made available to the user in different ways. Below are listed the files that are stored on the magnetic tape. Files 1 - 5 are in ASCII code with a block length of 7920 characters corresponding to a page with 60 lines, each containing 132 characters. File 6 is in HP format and is a direct copy of my own disc at the time of submitting the catalogue. This file is not aimed for printing but rather for computer-controlled investigations of cluster data.

TAPE FILES

FILE 1	INTRODUCTION DESCRIPTION	0.02 Mbyte
FILE 2	FORMATTED CATALOGUE	0.95
FILE 3	REFERENCES	0.05
FILE 4	ALIAS NAMES	0.21
FILE 5	TRACED TABLES	0.35
FILE 6	CATALOGUE IN HP FORMAT	1.21

II. COMPUTER CONTROLLABLE FILE

So far I have only described what could have been a printed catalogue, albeit one for which new editions can be frequent. If that were all that could be achieved with a computer, it would not be worth the effort. The far more important aspect is that it is possible to interact with the catalogue, changing parameters, deleting entries and adding new information. Thus, while a printed catalogue is ageing from the time of its submission, a computer-based catalogue is growing and changing continuously. Other advantages are the flexibility of list structures (Davis, 1977), the large storage capacity and the mobility of the data bank.

It has been my aim to create the sort of structure for the catalogue of open cluster data. Below I have summed up the advantages and the problems encountered so far. The advantages are obvious and I hope to show their value elsewhere, in connection with studies of galactic structure.

COMPUTER-BASED VERSUS PRINTED CATALOGUE

ADVANTAGES	PROBLEMS
Ease of revision	Inhomogeneity
Direct access	Notation of imprecision
Error removal	Inflexible formatting
Flexible ordering	Program errors
Large storage	Unfamiliarity
Mobility	

III. PROBLEMS

In the present context the problems are of interest. The most fundamental difficulty with all large collections of data values is the inhomogeneity. A data value may be different depending on telescope, exposure time or other observing conditions. It may also refer to a different interpretation of observed object. In a printed catalogue such details may be remarked on. Often such notes are, however, not made and then the printed catalogue has no advantage over the computer-based one. In the Prague-Budapest catalogue all published values are given together with their references. This may sometimes be desirable but it can be difficult for an investigator who is not familiar with the astronomical literature. I have for many parameters tried to select one single value for each cluster. Such a selection is necessary for applications but it may still suffer from a certain degree of inhomogeneity.

Connected with this problem is the one of assigning imprecision. An example: the radial velocity of a cluster is listed as one value, expressed in km/s. It may be obtained as a mean value of twenty accurately measured radial velocities or refer to a rough measurement of one possible member star. Yet, what we enter into the data base is one value per cluster. In this as in many other cases there ought to be some standard way of denoting quality of data. The printed catalogue can use a number of methods like colons, italics, underlining or variable number of decimals to give the reader a guide to his selection of data. In the computer-based catalogue such a function must be filled by a separate parameter giving weights or assigned values of imprecision.

Let me in this connection also make a comment about the limitations of format imposed by computers. If the colour excess of a cluster is $0\overset{m}{.}12$ I want to list that value. Another cluster has the less well known colour excess of $0\overset{m}{.}1$; this will appear as $0\overset{m}{.}10$ if the format (F4.2) is used. To beat the system I then decide to use free field format and the colour excesses are given as $0\overset{m}{.}12$ and $0\overset{m}{.}1$. All very well, but what about the cluster for which accurate photometry gave $0\overset{m}{.}10$? That also will be listed as $0\overset{m}{.}1$.

IV. CONCLUSIONS

Let me sum up by pointing out that computers have given us entirely new ways of comparing data and of drawing conclusions. However, to know the extent of possible bias and unreliability we must be in full control of the procedures. At the present, there appears to be no more efficient way to interrelate the large amounts of available observations than to use computer-based data banks. Thus it is imperative to develop good standards for critical inspection of the data.

References

Davis, M.S. 1977: In Jaschek, Wilkins: Compilation, Critical Evaluation and Distribution of Stellar Data. IAU Colloquium No. 35, p. 3.

Lyngå, G. 1981: Astronomical Data Center Bulletin, Vol. 2. In press.

CATALOGUE BIBLIOGRAPHIQUE DE VITESSES RADIALES STELLAIRES

M. Barbier et M. Petit
Observatoire de Marseille, France

SUMMARY

The continuation of the Abt-Biggs catalog of stellar radial velocities from 1970 is presented. A first part (6000 references) has been published on microfiches (CDS, 1980). Five thousand new references should be published in a near future.

Ce catalogue fait suite au catalogue de H.A. Abt et E.S. Biggs (1972). Nous avons commencé la bibliographie en 1970, date à laquelle s'arrête ce dernier. Il est mis à jour régulièrement au fur et à mesure de la parution des publications. Une première partie comprenant environ 6000 références de 1970 à 1975 est publiée sur microfiches (C.D.S. 1980). Elle n'est cependant pas exhaustive. Depuis deux ans 3000 références ont été perforées comprenant celles qui n'avaient pas été incluses dans la première publication. 2000 nouvelles références sont actuellement en cours de perforation pour 1980 et 1981.

Les principales revues susceptibles de publier de nouvelles mesures de vitesses radiales stellaires sont consultées.
En voici la liste avec leur abréviation dans le catalogue :

AA	Acta Astronomica
AAP	Astronomy and Astrophysics
AAPS	Astronomy and Astrophysics Supplement
AFA	Arkiv für Astrophysik
AJ	Astronomical Journal
AP	Astrophysics
AN	Astronomische Nachrichten
APJ	Astrophysical Journal
APJS	Astrophysical Journal Supplement
ASS	Astrophysics and Space Science
AT	Annals of the Tokyo Astronomical Observatory
BAIC	Bulletin Astronomical Institut Czechoslovaque

C. Jaschek and W. Heintz (eds.), Automated Data Retrieval in Astronomy, 293–295.
Copyright © 1982 by D. Reidel Publishing Company.

BAOB	Bulletin Astronomique de l'Observatoire Royal de Belgique
BOAB	Bulletin de l'Observatoire Astronomique de Belgrade
CAO	Publication of the Crimée Astr. Observ.
CRAS	Comptes-Rendus de l'Académie des Sciences - Paris -
DAO	Publications of the Dominion Astrophysical Observatory
IAJ	Irish Astronomical Journal
IBVS	Information Bulletin on Variable Stars
JO	Journal des Observateurs
LOB	Lick Observatory Bulletin
MAKL	Medelelingen van. het Astronomisch Institut van de Katholieke Universiteit Leuven
MEM	Memoirs of the Royal Astronomical Society
MESO	The Messenger (E.S.O.)
MN	Monthly Notices of the Royal Astronomical Society
MSAI	Memorie della Societa Astronomica Italiana
OBS	The Observatory
PASJ	Publication of the Astron. Society of Japan
PASP	Publication of the Astron. Society of Pacific
PAUC	Universidad de Chile - Departemento d'Astronomica - Separatas
PDDO	Publication of the David Dunlap Observatory
POAS	Publication de l'Observatoire Astron. de Strasbourg
POCC	Publication de l'Observatoire de Cerro Calan
POGB	Publication de l'Observatoire de Genève - Série B
POHP	Publication de l'Observatoire de Haute Provence
RASC	Journal of the Royal Astronomical Society Canada
RMAA	Revue mexicaine d'Astronomie et d'Astrophysique
ROB	Royal Observatory Bulletin
SAL	Soviet Astronomy - Letters
SAOC	South African Astronomical Observatory circulars
VA	Vistas in Astronomy

DESCRIPTION du CATALOGUE

Colonne 1 : numéro HD (Henry Draper catalogue)
Colonne 2 : numéros BD (Bonner Durchmusterung), C.D. (Cordoba Durchmusterung) ou C.P.D. (Cape photographic Durchmusterung)
Colonne 3 : numéro G.C.R.V. (General catalogue of radial velocities; Wilson, 1953)
Colonne 4 : identification dans d'autres catalogues ou listes
Colonne 5 : ascension droite et déclinaison pour 1950
Colonnes 6 et 7 : magnitude visuelle et type spectral publiés par l'auteur
Colonne 8 : vitesse radiale, qualité et nombre de mesures
Colonnes 9, 10, 11 : remarques, dispersions et références

Une description plus détaillée figure au début du catalogue.

REMERCIEMENTS

Les auteurs tiennent à remercier très vivement M. le Professeur C. Jaschek, Directeur du C.D.S., M. F. Ochsenbein, astronome au C.D.S. ainsi que tout le personnel du centre pour l'aide efficace qu'ils ont apportée et qui a permis l'édition de la première publication.

A Mme A. Viale, technicienne à l'Observatoire de Marseille pour sa précieuse collaboration pendant les deux premières années et à Mme M. Gerbal, technicienne à l'Observatoire de Marseille qui assure depuis deux ans avec grand soin le travail de perforation des données.

REMARQUE

Il est pratiquement inévitable dans un tel catalogue que des erreurs demeurent. Les auteurs seraient très heureux qu'elles leur soient signalées afin d'apporter les corrections nécessaires lors de la prochaine publication.

BIBLIOGRAPHIE

Abt, H.A., Biggs, E.S.: 1972, "Bibliography of Stellar Radial Velocities", Kitt Peak Nat. Obs.

Wilson, R.F.: 1953, "General Catalogue of Stellar Radial Velocities" Carnegie Inst. Washington, Publ. 601.

THE NEW CATALOGUE OF OPTICAL HII-REGIONS

P. Polechova
Observatory Petrin, Prague, Czechoslovakia

ABSTRACT

The principles of the work on the new catalogue of the optical HII-regions are described.

On the basis of thirteen catalogues and lists of the optical HII-regions the catalogue of these objects had been compiled by me (Marsalkova, 1974). The main source of the objects on the northern sky was the catalogue of Sharpless (1959) which up to now is the only one in computer-readable form, the southern sky being characterized by a larger number of smaller catalogues represented in first place by that of Rodgers et al (1960) and that of Gum (1955). For this already published catalogue (now referred to as the "old" one) only optical data were taken into account. The catalogue was supplemented by a list of excluded HII-regions like supernova remnants, planetary nebulae and galaxies.

One of the gains of the old catalogue was the comparison of various subdivisions of the nebular complexes published by the individual authors. E.g. the Cygnus X-regions was catalogued by Dickel et al.(1969) as 193 objects while Sharpless (1959) lists one object in this region only. Byt this was one of the most simple cases to solve.

The registration of all the subdivisions of the complexes listed by various authors was provided as a mean of following those authors whose description of the objects in the area was the most detailed one. This will not be precisely the case in the new catalogue, as will be mentioned later on.

The old catalogue removed a number of mistakes which appeared in the source catalogues mainly in the coordinates. One of them was the systematic error of the coordinates in the catalogue of Sharpless (1959) caused by the 20-years precession. It is interesting to point our just at this colloquium that this error (the origin of which was explained in

C. Jaschek and W. Heintz (eds.), Automated Data Retrieval in Astronomy, 297–300.
Copyright © 1982 by D. Reidel Publishing Company.

the old catalogue - Marsalkova, 1975, p.10) could never have been revealed by a computer.

The imperfections of the old catalogue were the following.

It was not published in computer-readable form.

Running numbers were used instead of some system of numbering open to later discoveries. This was because of the frequent cross-references included and because of the visibility of the entries on the page.

The old catalogue was inhomogeneous : since Lynga's atlas (1972) shows only the objects of higher brightness and covers the narrow band around the galactic equator, the scale of the most detailed maps used for the southern sky (Rodgers et al., 1960) is 17 times worse than that of the POSS, which was the basis for the cataloguing of the objects on the northern sky.

The scale of POSS was not well used : e.g. the nebulae M 42 and M 43 were listed in the catalogue of Sharpless (and therefore also in the old catalogue) under the common name S 281.

Contrary to this, the detailed subdivisions of the objects according to e.g. Dottori and Carranza (1971) should not have been regarded as the representative ones because of the low angular resolution (110'/mm) of their survey.

In the new catalogue we shall try our best to remove or suppress all these defects.

Inhomogenity will be supressed by the use of the coordinate boxes referred to in my previous paper (Polechova, 1982). It will be possible to incorporate the data into the body of the catalogue without changing the numbers of the objects. The objects smaller than 6' will not be catalogued in the southern sky until the ESO/SRC red maps become available ; this implies that the two-decimal digit range of the names in the system of the catalogue will not be covered up to that time.

The system of boxes also exludes those unjustified subdivisions, because instead of following the author whose subdivision of the complex is the most detailed one, the confrontation with the radio maps will be essential.

The confrontation of the radio maps with the optical maps is appropriate also because of the fact that the nature of some non-radio emitting objects catalogued as HII-regions appears to be doubtful. The importance of the mutual confrontation of the optical and radio results for the purpose of the catalogue of HII-regions shows up for instance in the example of the object S 88 : the position of the object named S 88 in the paper of Silverglate and Terzian (1978) indicates that the radio measurements reported in their paper do not concern the true S 88 object. (Fig. 1).

Some work for the new catalogue has been already done. At present a majority of the radio continuum maps of HII-regions in the different frequencies and different HPBW's has been assembled from about thirty papers in the scale of the POSS.

I have had no constant collaborator during this phase of the work, except some young people associated with the Petrin observatory which helped me occasionally with the preparation of the material for the abovementioned comparison work. I am specially grateful to J. Arnoltova, K. Danihelka, J. Holanek, J. Kaspar, P. Kasparek; S. Lego, J. Manek, J. Soukupova and to my colleague T. Netopil.

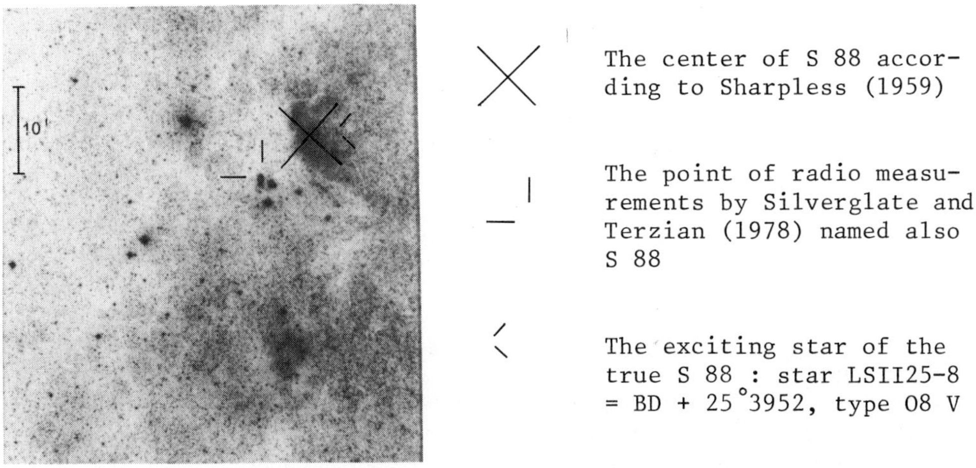

Fig. 1 The reproduction of the POSS red map with the object S 88 (α_{1950} = 19 h 43.9m, δ_{1950} = -25° 13'). See also the text at the top of this page.

REFERENCES

Altenhoff W.J., Downes D., Pauls T., Schraml J. : 1979, Astron. Astrophys. Suppl. 35,23
Angerhofer P.E., Becker R.H., Kundu M.K. : 1977, Astron. Astrophys. 55,11
Colley D. : 1980, Monthly Notices Roy. Astron. Soc. 193, 495
Colley D. : 1980, Monthly Notices Roy. Astron. Soc. 192, 377
Day G.A., Warne W.G., Cooke D.J. : 1970, Aust. J. Phys. Astrophys. Suppl. N° 13,11
Dickel H.R., Wendker H., Bieritz J.H. : 1969, Astron. Astrophys.1, 270
Dottori H., Carranza G. : 1971, Astrophys. Space Sci. 13, 180

Fanti C., Mantovani F., Tomasi P. : 1981, Astron. Astrophys. Suppl. 43,1
Felli M., Churchwell E. : 1972, Astron. Astrophys. Suppl. 5, 369
Felli M., Habing H.J., Israel F.P. : 1977, Astron. Astrophys. 59, 43
Felli M., Perinotto M. : 1974, Astrophys. Space Sci. 26,115
Goss W.M., Day G.A. : 1970, Aust. J. Phys. Astrophys. Suppl. N° 13,3
Goss W.M., Shaver P.A. : 1970, Aust. J. Phys. Astrophys. Suppl. N°14, 1
Gum C.S. : 1955, Memoirs Roy. Astron. Soc. 67, 155
Harten R., Felli M. : 1980, Astron. Astrophys. 89, 140
Haynes R.F., Caswell J.L., Simons L.W.J. : 1978, Aust. J. Phys. Suppl. N° 45, 1
Hughes V.A., Viner M.R., Kidd A.M. : 1977, Monthly Notices Roy. Astron. Soc. 179, 525
Israel F.P. : 1973, Astron. Astrophys. 27, 143
Israel F.P. : 1976, Astron. Astrophys. 52, 175
Israel F.P. : 1977, Astron. Astrophys. 59, 27
Israel F.P. : 1977, Astron. Astrophys. 60, 233
Israel F.P. : 1977, Astron. Astrophys. 61, 377
Israel F.P. : 1978, Astron. Astrophys. 63, 328
Kallas E., Reich W. : 1980, Astron. Astrophys. Suppl. 42, 227
Lynga G., Hansson N. : 1972, Astron. Astrophys. Suppl. 6, 327
Marsalkova P. (Polechova P.) : 1974, Astrophys. Space Sci. 27, 3
Matthews H.E., Goss W.M., Winnberg A., Habing H.J. : 1977, Astron. Astrophys. 6, 261
Mezger P.G., Henderson A.P. : 1967, Astrophys. J. 147, 471
Mezger P.G., Höglund B. : 1967, Astrophys. J. 147, 490
Polechova P. : 1979, Ph. D. Thesis, Charles University, Praha
Polechova P. : 1982, this volume, p. 207.
Retallack D.S., Goss W.M. : 1980, Monthly Notices Roy. Astron. Soc. 193, 261
Rodgers A.W., Campbell C.T., Whiteoak J.B. : 1960, Monthly Notices Roy. Astron. Soc. 121, 103
Schraml J., Mezger P.G. : 1964, Astrophys. J. 156, 269
Sharpless S. : 1953, Astrophys. J. 118, 362
Sharpless S. : 1959, Astrophys. J. Suppl. 4, 257
Shaver P.A., Goss W.M. : 1970, Aust. J. Phys. Astrophys. Suppl. 14, 77
Silverglate P.R., Terzian Y. : 1978, Astrophys. J. 224, 437

CATALOGUE OF PLANETARY NEBULAE AND THEIR NUCLEI

Acker A., Marcout J.
Observatoire de Strasbourg - France

Since the catalogue by Perek and Kohoutek (PK) has been printed in 1967, numerous observations and theoretical studies have been made, and new planetary nebulae (NP) have been discovered, in addition to the 1036 PN indexed by PK. We have compiled a bibliographic index from 1965, covering about 80 papers per year ; a catalogue of 1455 PN and a bibliographic index have been established (Acker, Marcout, Ochsenbein, 1981).

Now, two important catalogues are being compiled which result from the joint work of the Observatories of Strasbourg, Bordeaux, Marseille, the Laboratory of Astronomy of Montpellier, and the Observatory of Haute Provence.

1. CATALOGUE OF CENTRAL STARS OF PLANETARY NEBULAE

This catalogue concerns about 250 nuclei of PN (NPN) and will be established by A. Acker (Strasbourg), M. Chopinet (Bordeaux), F. Gleizes (Montpellier), J. Marcout (Strasbourg), with the help of F. Ochsenbein (Centre de Données Stellaires de Strasbourg) and J.M. Roques (Montpellier).

1.1. The data of observation published since 1965, as well as the bibliographical sources are presented :

- designations : the names HD, BD, CPD, ... of about a hundred NPN are given ; for all of them, the PK n° and the usual name of the PN are indicated.
- coordinates : the values α and δ are given for 1950, 1985 and 2000, values which are accurate ($\pm$ 1" to $\pm$ 2") for about a hundred bright NPN, and less accurate for the others ($\pm$ 0$'$.5)
- magnitudes : the available values m_p, m_{pv}, U, B, V are indicated.
- spectral type : for about 120 NPN, a spectral classification has been

C. Jaschek and W. Heintz (eds.), Automated Data Retrieval in Astronomy, 301–303.
Copyright © 1982 by D. Reidel Publishing Company.

UNIVERSITY COLLEGE LIBRARY CARDIFF

given, although the types are usually ill-defined. Most of them correspond to hot stars (continuous, WR, O, Of, ...) ; except twenty stars (A, F, G) which are either variables or binaries.

- radial velocities : we give the radial velocity of the star, the one of the PN, and the velocities of expansion, from the study of stellar winds (for the PN studied with IUE in particular) and from the splitting of the nebular lines.

- particularities : the data concerning the variability and the binarity (visual, spectroscopic, eclipsing) are indicated.

1.2. The bibliographic references are collected. We give the authors, year, title of the paper, name of the magazine, volume and page numbers, for the calculation of distances, luminosities, temperatures and masses in connection with the type of population and the age.

1.3. Field-maps are provided for every NPN

- BD map - field : 1.5° x 1.5°
- map from Palomar - field : 15' x 15'

This catalogue will be available for IAU Symposium N° 103 "Planetary Nebulae", London, August 9 - 13, 1982.

MONOCHROMATIC ATLAS AND CATALOGUE OF LARGE PLANETARY NEBULAE

This atlas, started in 1981, will present monochromatic images and bibliographic information for about thirty of the brightest PN with a diameter $\emptyset \gtrsim 30''$. It will be prepared by A. Acker (Strasbourg), M. Chopinet (Bordeaux), C. Hua (Marseille), R. Louise (Marseille), with the help of J. Marcout and F. Ochsenbein (Centre de Données Stellaires, Strasbourg) and is to be published as a book in 1983-1984.

2.1. Monochromatic images are being taken by R. Louise using the 120 cm telescope of the Observatory of Haute Provence (O.H.P.) with [OI] , [OII] [OIII] , Hα, Hβ and [NII] filters. These photographs give us indications about the morphology, in particular about the presence of external envelopes and halos. They enable as well to study the variations of intensity of the emission for these elements, at various points of the PN (resolution about $5''^2$).

Spectra of these PN are taken by A. Acker and M. Chopinet with the "Multiphot" system set up at the Coudé focus of the 152 cm. The resolution is compatible with the one of the photographs ($5''^2$) for an exposure inferior or equal to 10 minutes. Two ranges are used, each one 600 A wide, centered on 4800 Å and 6500 Å. We will determine the variations of the ratios of line intensities at various points of the

PN ; these measurements will be connected to the previous ones.

For the most extensive PN (listed by Abell) images are obtained by C. Hua with the I.P.C.S. system set up at the Cassegrain focus of the 193 cm telescope of the O.H.P. ; these tracings give immediate quantitative information.

2.2. Observational data and bibliographic references are provided for all these objects : designations, coordinates (1950, 1985, 2000), sizes, magnitudes, morphological type, excitation class, radial velocities and velocities of expansion, magnitude and spectral type of the nucleus.

These catalogues summarize what we know at present. They will permit to direct future observations, and to bring out certain general properties of the PN and their nuclei.

REFERENCES

Acker A., Marcout J. and Ochsenbein F. : 1981, Astronomy and Astrophys. Suppl. Ser. 43, 265

Perek L., Kohoutek L. : 1967, Catalogue of Planetary Nebulae, Czechoslovak Akad. of Sciences

NUMERICAL DATA DOCUMENTATION IN PHYSICS IN THE FEDERAL REPUBLIC OF GERMANY

G. Ebel
Fachinformationszentrum Energie, Physik, Mathematik GmbH
Karlsruhe, Federal Republic of Germany

Owing to the ever-increasing number of publications it has become more and more difficult for scientists to find physics data relevant to their respective activities in the primary scientific and technical literature. With the exception of their special field of research they have to rely on data compilations. In many fields of physics, however, such data compilations do not exist or do not fulfill the requirements of the users, either because they are out of date or because they are published in places that are not well known. This unsatisfactory situation has been confirmed by an inquiry performed by the German Physical Society in 1972.

In 1974 the German Federal Government passed the "Program of the Federal Authority for the Promotion of Information and Documentation". The aim of this program was the establishment of Specialized Information Centers for the different fields of activity of modern society, such as physics, chemistry, life sciences, earth sciences, materials science, engineering, social sciences, humanities, legislation, transportation etc. A number of these information centers have come into existence during the last years, amongst them the Fachinformationszentrum Energie, Physik, Mathematik in Karlsruhe. It is acting as the central scientific and technical information and documentation center for the fields of
- energy research and technology (nuclear and nonnuclear)
- physics
- mathematics
- astronomy and astrophysics
- space research, aeronautics and astronautics.

The Fachinformationszentrum has a staff of roughly 23o people about 6o of them being scientists. All activities are supported by modern information technologies (including two SIEMENS computers 776o and 7541) which provide comprehensive and efficient information services on scientific and technical literature, numerical data and on-going research projects. In the fields of astrophysics and space sciences three databases for the retrieval of the relevant literature are offered

C. Jaschek and W. Heintz (eds.), Automated Data Retrieval in Astronomy, 305–308.
Copyright © 1982 by D. Reidel Publishing Company.

- Physics Briefs
- International Nuclear Information System (INIS)
- NASA/ESA Database

A coordination with the work of the Astronomisches Recheninstitut in Heidelberg for the Astronomy and Astrophysics Abstracts is envisaged.

Since the beginning of 1979 the databases of the Fachinformationszentrum are accessible online via the public telephone network and since January 198o via EURONET. This network, set up by the member countries of the European Communities transmits information using the latest packet-switching technologies. Within the next several months a number of national packet-switching data networks will be connected to each other, such as DATEX-P in Germany, TRANSPAC in France and TELENET-TYMNET in USA. This will lead to a lower priced and more comfortable direct access to databases in different countries.

In the field of numerical data documentation Germany has a long tradition in the publication of comprehensive handbooks in physics and chemistry. The three most widely known of these are:

- Landolt-Börnstein Numerical Data and Functional Relationships in Science and Technology (6th Edition and New Series)
- Gmelin Handbook of Inorganic Chemistry (8th Edition)
- Beilstein Handbook of Organic Chemistry (4th Edition)

Each year a large number of new volumes is added to these handbooks.

The Fachinformationszentrum takes into account the rising demand for selected and evaluated data by gradually expanding data documentation and information. It acts as a national information center for physical data within the Federal Republic of Germany and cooperates with other national and international organizations. In this respect the cooperation with the International Atomic Agency in Vienna for the fields of nuclear data and for atomic and molecular data and with CODATA, the ICSU Comittee on Data for Science and Technology, should be specially mentioned.

The center's own activities are focused on the following special subjects:

- high energy physics
- atomic and molecular physics
- crystallography
- nuclear physics
- solid state physics

Data documentation in other fields of physics as well as on properties of materials is being prepared. Expansion into the energy research sector has been started.

Critical selection, evaluation and compilation of data is carried out in close cooperation with research groups at universities, research centers, Max-Planck-Institutes etc. Most of this work is published in Physikdaten/Physics Data, a series establiahed by the Fachinformationszentrum for the publication of data compilations in selected fields of physics. Up to now 3o issues have been published in this series. Most of the compilations are either constantly brought up to date or else replaced by new editions in due time.

Special reference should be given to a bibliography of data compilations in physics. This bibliography has also appeared in the Physics Data series and contains about 3ooo existing data compilations from all fields of physics on a worldwide basis. Nearly all of these data compilations are available in a reference library at the Fachinformationszentrum.

In addition to printed data compilations which have a long tradition in physics computerized data bases have been established in more recent years. The reasons are that large amount of data can be handled conveniently only by computer and that in many cases these data are required for further computer calculations so that they must be available in computer readable form anyhow. In the last part I will give a survey on the larger numerical databanks in physics that are available in Germany.

The Fachinformationszentrum offers the following numerical data banks that have mostly been established in collaboration with other institutions:

- Karlsruhe Charged Particle Reaction Data (KACHAPAG). Cross sections and other data for nuclear reactions with charged particles.
- Cambridge Crystallographic Data Files. Crystal structures of organic and organometallic compounds.
- Inorganic Crystallographic Data Bank. Crystal structures of inorganic compounds.
- Data Bank on Electron Densities in Crystals. Electron density distributions in crystals.
- Data Bank on Crystal Structure Types. Systematic relationships between crystal structures.
- Carbon - 13 Nuclear Magnetic Resonance Data Bank. Nuclear magnetic resonance spectra of organic compounds.
- Data Bank for Molecular Physics in the Gaseous Phase. This data bank is in preparation at the Section for Structure Documentation of the University of Ulm and the Computer Center of the University of Freiburg. It will give a detailed access to all molecular constants for inorganic and organic molecules published in the literature.

All these databanks will be handled by the general database management system ADABAS and will be made available in online access via the public data networks.

In addition to the Fachinformationszentrum there are a number of other institutions at which numerical data banks are available such as

- Kernforschungszentrum Karlsruhe: Nuclear Data Bank Karlsruhe (KEDAK)
- Deutsche Gesellschaft für chemisches Apparatewesen (DECHEMA) in Frankfurt: DECHEMA Property Data Bank for Chemical Substances
- Betriebsforschungsinstitut des Vereins Deutscher Eisenhüttenleute

in Düsseldorf: Data Bank on Steels and Ferrous Materials
- Lehrstuhl für Metallurgie der Kernbrennstoffe und Theoretische Hüttenkunde at the Technische Hochschule Aachen: Thermochemical Data Bank on Inorganic Materials.

As a final statement one may say that a great demand is faced with an unsatisfactory offer. The number of numerical data banks will continue to increase in the future but not as fast as expected due to the high costs of compilation. These high costs result from the fact that the quality of the stored data must be guaranteed, i.e. they must be checked by time-consuming intellectual evaluation. This means that in physics a situation where all relevant numerical data will be stored for retrieval in special data banks will not be arrived at for decades to come.

FINAL COMMENT

B. Hauck
Institut d'Astronomie de l'Université de Lausanne
CH-1290 Chavannes-des-Bois, Switzerland

During the General Assembly of the IAU at Brighton in 1970, the idea of a stellar data centre was received with a fair amount of scepticism. This colloquium shows clearly that this scepticism was an error of appreciation. We see today that major data centres exist, here at Strasbourg, in Japan, in the USA and USSR, and that a close collaboration exists between them. In addition, many groups in various fields bring very valuable contributions. The development of astronomical networks is also a pleasant thing to mention; but it is certainly less pleasant, as regards ground-based data, to speak about funding. Governments or scientific authorities give money for very expensive instruments (e.g. large telescopes) with which it is possible to obtain data - but expensive data. These same authorities do not pay enough attention to the preservation and dissemination of these data. This situation is certainly better in space astronomy, which is mainly due to the existence of the NSSDC. An effort should be made in two directions: facilities to publish data and extension of financial support to data centres.

C. Jaschek and W. Heintz (eds.), Automated Data Retrieval in Astronomy, 309.
Copyright © 1982 by D. Reidel Publishing Company.

REPORT OF DISCUSSIONS

Wulff D. Heintz
Swarthmore College, Swarthmore, U.S.A.

A. EXISTING DATA CENTERS

In connection with the paper by Mead, videodisks were mentioned as a recommended storage medium, as they have a long lifetime, can store graphs, and are inexpensive to reproduce. Vette commented that magnetic tape now available also is quite durable.

Heck responded to the question how the data will be preserved: Several copies of the IUE images on magnetic tape are presently maintained around the world: Two or three by NASA at different centers, one at Rutherford and Appleton Laboratories, and one at Vilspa. Each agency is responsible for finding the best way to preserve its data set. Lyngå noted that the information storage on microfiche is very efficient from the viewpoints of packing density and of durability, but the major difficulty currently is in the automatic film reading, for which a technique need be developed. Vette confirmed that no reliable method is yet available although work on this potentially important method progresses. Davis pointed out that the creation of large quantities of machine readable observational data is not limited to spaceborne instruments. For instance, many of the instruments now being used at the Mount Hopkins Observatory of the Center for Astrophysics record on magnetic tape, accomodating the annually growing rate of produced data. It was also noted that the astronomical storage needs are not extreme. Even the data from the Space Telescope will be less than what weather and Earth satellites supply.

Schmitz added to his presentation on the cataloguing of infrared objects that coordinates are assigned from original sources as referenced, or else by measuring approximate positions from published maps. When working in the near-infrared (690 to 890 nm), McCarthy noted that many of the positions are already found in Dearborn Observatory catalogs, some also from Case Observatory and in ApJ 242, 938 (1980). Westerhout inquired about the recording of extended and multi-peaked sources. Schmitz will provide maps for these 10% of the catalog entries.

C. Jaschek and W. Heintz (eds.), Automated Data Retrieval in Astronomy, 311–322.
Copyright © 1982 by D. Reidel Publishing Company.

Westerhout (responding to an inquiry by Vette): The most accurate method of determining polar motion is currently from the averages obtained by the Bureau International de l'Heure (BIH), this being the only "standard" reference system at hand. However, at the level of consistency of the BIH data, both connected-element interferometry and the laser ranging data show systematic trends which are similar. Probably the most precise method is Radio VLBI, but nothing is known yet about its systematic effects.

Comments resulting from the discussion of the Sedmak/Santin paper on ASTRONET: Presently only the system located in Bologna is in operation. National and international links are planned to be added as soon as the other systems will become operative. The network is funded by the participating institutions and from a grant from the Government of Italy. Two or three simultaneous graphics users are anticipated, and the software standards are not yet defined.

Discussion following the paper by Locke centered on the availability of data in form of published catalogs. Concerning the possibility to combine data from various satellites, even from different wavelength langes (Heck): Yes, it is intended to compile the data, as they become available from different satellites, and to tag them according to different parameters, such as source, wavelength, specific time, and others as may be of interest to requesters. Schmitz inquires if the NSSDC agreement includes classified data. Locke: All users are encouraged to submit their data; several projects have been unclassified and sent to the Center. McCarthy commends the colleagues at NASA, NOAA, and NSSDC for efficient data accessibility, in his case connected with the weather survey of a new observatory site. The material was usually readily obtained, or at least the reference where to get it.

The Faint Object Camera was the subject of Macchetto/Perryman. What is the cosmic-ray background rate (Vette) ? The effect of low-energy electrons and protons will be minimised by shielding. Certain other events will be recognised as particle-induced by the pattern-recognition unit. Events that escape this recognition will, however, give rise to one count only in the final image. Figures are available from NASA on the particle-event rate which is a function of the orbital position. Like other Space Telescope instruments, the FOC will be switched off during periods of high background rate. - What is the status of projects on the processing of data to come from the HIPPARCOS astrometric satellite (Heck) ? The Announcement of Opportunity for offers to reduce all data gathered through the lifetime of HIPPARCOS has been recently released by ESA. Two independent consortia will be selected, and all of the data be forwarded to them. The reduction is expected to result in a final astrometric catalog, containing positions, proper motions, and parallaxes for all program stars, perhaps two or three years after termination of the mission. A more complete overview of the scientific coordination of the HIPPARCOS project (launch expected in 1986) has been published by Perryman in the Bulletin d'information CDS Strasbourg, no. 21, p.40 (1981).

In the discussion following Terashita's presentation, Mead described the astronomical database retrieval at NASA-GSFC: The basic retrieval program keeps the catalogs in their original format. For each individual catalog the computer program contains these formats and retrieves the entire entries based on catalog identification numbers. This was found to be more satisfactory than rearranging catalogs to fit some standard format, in particular because all catalogs differ with respect to arrangement and contents. Thus compatibility is hard to achieve if it is desired to retain all of the catalog data.

B. DATA NETWORKS IN ASTRONOMY

The report on EURONET led to the question whether connections with different national networks are expected. Salle feels that this appears to be doubtful. EURONET can be considered as the triggering factor toward a joint European network, and its future will depend on the success of the participants in linking their networks together.

De Pablo had this comment to the paper by Carnochan: An excessive proliferation of private or domestic networks for the astronomical data interchange seems to be evident. The solution of the particular interlinking problems is effort and time consuming. Quite often the adopted solution is to take methods, protocols, and even a specific hardware model of one manufacturer. Promotion of common methods of data exchange and software production could be one of the tasks of IAU in the future. Such methods should be manufacturer independent and internationally acceptable. As Wilkins explained, one of the reasons for the decision to link the six computers on STARLINK was to encourage the development of common software. Information about new software, as well as the programs themselves, are easily passed over the links. Much more cooperation and less duplication appear to have resulted than would have been the case with independently installed computer systems. Subsequent discussion showed serious problems to result from the software incompatibility between various systems. The question if the calling of programs at various places would jam lines (Heck), was answered by Carnochan: Sending information to and from several machines simultaneously will not jam lines since the system connecting the computers will be able to avoid that. But certainly the low speed of the lines does mean a serious degrading of response times once a certain amount of traffic and of users is exceeded.

C. NEW HARDWARE

Perryman comments on the high recording speed offered by holographic video systems. In particular, it would be quite useful to record digital speckle images in this way. For example, 50 frames of a 500 x 500 x 16 bit photon-counting system per second could be recorded digitally with such a system, and this corresponds to a recording rate of 25 megabytes/s.

D. RECENT SOFTWARE DEVELOPMENTS

Albrecht emphasized that the software production for the Space Telescope is deadlined. Thus he cannot afford to wait for a standard format, such as that being developed for STARLINK, and mentioned by Wilkins. In response to Terlouw, Albrecht explained that the Virtual Array approach is also for machines without mapping hardware; the image is mapped by software. This is certainly less efficient, and cannot be made as efficient as would be possible in theory, because of the trade-off between speed and user-friendliness. Should a program using Virtual Arrays be found to lack the speed perhaps necessary for certain purposes, some parts can be replaced. The question arose (Polechova) if the conversion of photographic data to tape or disk was limited to plates, excluding copies like the Palomar Survey prints. Albrecht has for the PDS microdensitometer an "opaque sample option" available. It can also be done with a video camera, since positional accuracy is not that important, especially when paper prints are digitized. Such video digitizers operate already in quite a few institutions, and should be common in the near future as their prices diminish.

McCarthy agrees with Eichhorn's discussion of positional precision and accuracy, and would like to have heard the word "standards" also with respect to the photometric systems.

Bijaoui elaborated on the proposed handling of his - surely voluminous - catalog. At each request for a field the Data Center provides a tape. It is easy to update, thus maintaining completenesss of the catalog. Davis addressed the problem of extracting photometric information for stars and galaxies when they are comparable with, or even fainter than, the background. Bijaoui pointed out that a second problem goes with that, viz., the separation of objects in crowded fields. Many good algorithms have been developed for the latter task, but they will not work very well on too faint objects (m>20), compared with the background. Crowded fields require a good signal/noise ratio and flawless images.

Albrecht asked Mistrik about the suggested improvements in planning of the project life cycle, and in particular the advocated compiling requirements. This is fine if the user is an airline or a bank. In research the requirements analysis is very difficult, and the scheduling of a project which pushes the state-of-the-art is close to impossible. Mistrik agrees that to compile requirements is quite difficult for anyone but the researcher himself. A careful analyst would target certain areas to be questioned, and would divide the field into sub-areas which can be classified as generic, group specific, and special, in order to question at least in the generic area. Major decisions in the other areas are left to the researcher. In smaller research projects the researcher is better off compiling requirements himself, from practical viewpoints, although the analyst will be able to help through transforming them into precise specifications.

In response to several questions, Stein described the selected star catalog: It is essentially a subset of the SAOC and should therefore show the same star distribution and galactic concentration. Most positions and proper motions are directly from SAOC, a few from AGK3. The 5°x 5° fields do not have the same centers as those of the ESO/SERC survey. Double stars not resolved by the sensor are eliminated as they give an incorrect centroid. Therefore also the magnitude information (to indicate doubles) needed not be retained. The software generating the mission catalog and associated tasks, such as star retrieval from mission records, can later be made available (as Warren suggested); at this time the documentation is still incomplete. Carnochan described an automatic arrangement of several catalogs merged with the SAOC in strips of 1° of ecliptic longitude for a UV experiment.

Lavrov uses the term "table" as meaning "relation" in a relational database, and can readily separate tables for any object type.

Warren: The services of the Data Center are offered to the entire scientific community. The information is circulated through the Astronomical Data Center Bulletin, published as needed (semiannually or so); observatories and astronomy departments in the Americas and selected institutions worldwide are on the mailing list. Specific catalog subsets (mission catalogs) can be generated from detailed specifications, provided that the number of requests and the time needed to develop ad hoc software will not be excessive. Data are at intervals corrected for errors discovered by users. It was intended to inform individual users about errors in the material they had received; but this large job will have to be done through the Bulletin.

E. BIBLIOGRAPHIC SERVICES

Following his Invited Paper, Schiminovich states that category headings only have been devised so far while classification of papers has not begun. The periodical publication of the computer-generated index, suggested by Warren to aid authors in classifying their papers, would amount to the use of algorithmic techniques as tools for the design and support of classification systems, and would be a welcome development, short of complete implementation of the automatic techniques for the actual processing of the papers.

Commenting on the paper by Laloe, Jaschek asks that authors be encouraged to get their bibliography on tape as soon as possible. Longer lists of objects in publications have to be bodily referred to as "lists", though there is no consensus on the minimum number of objects which would make a list.

A machine readable version of the Astronomy and Astrophysics Abstracts (AAA) and the loading of their back issues onto bibliographic services was suggested, but is not available. The subject of key words received discussion here and at special Working Group meetings.

Collins: Efforts are made to add identifiers of objects besides coordinates, to list alternative spellings, and to recognise certain abbreviations. Papers may be scanned for additional designations and names not in the title, although for not more than 10 objects.

F. COPYRIGHT

Guerassimov cited his paper as an example for non-exclusive copyright: Although to be published by Reidel, it is work for UNESCO and thus may be used by anyone. It has not been decided whether computer programs are protected by copyright law or by patent law (question by Wilkins), though the majority opinion appears to prefer copyright law as being the most appropriate. Carnochan expects that the "economic" side of copyright will soon be sorted out. It is becoming economical for publishers to hold their journals in full text on computers. The scientist will be able to log into the computer and to receive immediately the article he requires. In this manner the publisher will receive the royalties which he does not get from photocopying papers. Such schemes are currently as inexpensive to operate as services such as the supply of requested papers by the British Library. One process whereby material is placed onto videodisks uses a filming technique. Thus scientists may want to be a bit careful when signing away the film rights to their work.

G. EDITORIAL POLICIES AND NOMENCLATURE

The discussion explored a variety of items addressed in the Invited Paper by Lequeux, in particular, in which way the traditional publication of data and their storage in centers might best complement one another. Jaschek felt that journals should not be charged with the publication of long catalogs, for instance, lists with more than 500 entries. These could go into observatory publications or directly into datacenters, with only a summary published in periodicals. Some IAU Commissions have advised against this policy. Worley - working in the production, compilation, and use of data - feels dissatisfied with the editorial treatment of data in several journals even now. The reading of older journals shows discussions and analyses which often are outdated, even appearing irrelevant and quite naive, yet together with data which still remain useful. Present editorial policy should not go toward continuing deemphasis of data in the journals, or else the short lifetime of papers will suggest an alternative way of reducing costs, i.e., to print on cheap newsprint stock - with the advantage that shelf space is saved as journal volumes self-destruct in a reasonable time.

To insist on all of astronomy in one journal seemed to Jaschek to be just a habit from history. The main journal of Astronomy and Astrophysics, for instance, could be split into specialised publications which cover smaller fields. Subscribers almost always use only a small fraction of the contents, but pay for all of it.

The problem of object designations can, Jaschek believes, be solved with the aid of the master list of catalog abbreviations prepared by Spite and co-authors, and nearing publication. Editors can refuse scripts containing obscure and unexplained designations if the referees do their work properly. Lequeux said that editors naturally cannot examine papers for compliance with IAU recommended nomenclature (for instance, the new denotation for clusters adopted recently by Commission 37, and mentioned by Lyngå); this is the reviewers' job. Jaschek also supports the trend toward camera-ready manuscripts. With the mean life of a paper of less than a decade, there is no point in much elevated expenses merely for a bit of esthetics. Eichhorn recommended the incentive of a reduction of page charges for camera-ready papers, instead of an accelerated publication.

Carnochan mentioned word processors as an alternative production method besides typesetting and camera-ready copy. Each article is essentially typed once only, including corrections and redrafting on the word processor, and can via computer linkup be used directly as input to the computer typesetter of the publisher. Since the paper is machine readable at all stages, abstracts can be early dispatched to the terminals of abstracting services, which saves time and work for them also.

One Working Group of IAU Commission 5 is preparing a vocabulary which should help toward unification of the key-word lists in use (Lantos). Another WG is trying to get a bit of homogeneity into the rather disintegrated policies sometimes imposed by publishing firms, and some of the subjects addressed in the extensive discussion (of which only a selection is reproduced here), following the paper by Lequeux, will have to be taken up again at the next IAU conference.

Following the suggestions in the paper by Spite and Ochsenbein, Mead emphasized that double designations - a long-standing IAU recommendation - are still very important since errors (and loss of the data) may arise in databases owing to typographical errors in single designations which cannot be checked. As reiterated by Commission 5 in 1979, coordinates are acceptable identifiers with a clearly specified equinox. Heck reported similar identifying requirements adopted by IUE, viz., 1950 coordinates plus an identifier, and if the latter is not from a major catalog ("exotic"), a written evidence from the literature. New running numbers are sometimes introduced in catalogs when there is really no need for them as identifiers; they merely complicate retrieval. The Catalogue of Stellar UV Fluxes used only HD or DM numbers or (in their absence) 1950 positions (Carnochan). The numbering of stars in clusters will be the subject of a Commission 37 resolution at the next General Assembly. Worley warned of problems which may additionally be introduced by standardized designations into an already confused field. Some codes are already ambiguous, and younger astronomers attempting to interpret the older literature will be confused by codes since eliminated. - Since the discussion on nomenclature arranged by Commission 5 in Montreal, some scientific commissions have taken up the matter, and the hoped-for coordination will be to the benefit of data efficiency.

The need for nomenclature rules, clear enough so that they can be applied by every astronomer, seemed to be agreed on. This was discussed at several past meetings like the present one, as Mead pointed out, and in order to start a standardized numerical nomenclature right away, it was high time to finalize such rules. Opinions were divided, however, whether identifiers be assigned by the individual astronomer or by an agency, such as a data center covering the subject area. The needs are that a name be unambiguous, be available as soon as possible, and be expected to become a permanent, useful identifier. It may be possible to wait for an official name while there is a preliminary designation, consisting of the discoverer's code and list number, and accompanied by the coordinates and the full description of the object, in order to avoid confusion and to clarify the type of object; as an exclusive identifier system, however, discovery list numbers are clearly unacceptable (Polechova). The risk in this procedure: Preliminary designations, intended to be temporary, will be perpetuated by inertia. The use of the Nomenclature Booklet by Lortet et al should help authors to avoid confusion. Warren urged that new catalogues be prepared using a homogenized, new designation system, but that every effort be made to cross-identify objects with previous designations; otherwise any new system would be impractible to use. Vette cautioned not to be too optimistic, from experience in information from artificial satellites; as long as there are various groups working on the same objects - spacecraft or stars - there will be aliases.

Additional difficulties arose with the nomenclature of extended objects, on which Polechova had reported. The size of the object serves as a second identifier, although not entirely independently as revised size measurements may modify the precision of the identifying position. Polechova stated that the information on the range of dimensions is simply and expediently a byproduct of the naming system; the information about frequencies is deemed necessary and, besides, already widely used. The cross-identifications of objects including corrections are considered the most useful advance in her already published catalog of H II regions, and this progress should be preserved (and extended to new designations) in the computer readable form of the catalog of extended objects. NGC 2024 is an example for gross confusion when names are assigned without well-defined agreement. The preference for galactic over other coordinates as identifiers received several supporting comments from the audience. Objects with uncertain positions, such as X ray sources, pose a problem similar to that of extended sources. Designations were revised between the 3U, 4U, and Einstein Observatory catalogs as positional accuracy improved. Dickel distinguished two modes of operation when observing the interstellar medium: (a) targets identifiable by types and (galactic) coordinates in surveys and catalogs; (b) studies of a given region such as the Orion Molecular Cloud. For the latter kind a different sort of designation is described in the Nomenclature Booklet (Lortet, Fernandez, and Spite) and may appear a bit cryptic, but again, there are two identifiers to be quoted. In general, one should not attempt to put too much of information into the identifiers, lest they may become impractical (Vette).

Fernandez asks that word be spread about the forthcoming compilation of existing catalog designations, expected to be published as a special Astr.Astrophys.Supplement. The authors have purported to follow the common procedures found in recent literature, and introduce new abbreviations only when there are already duplications. The request is made that, for any input sought to dictionaries and compilations, contact name, addresses, and input requirements be widely circulated.

The Catalog of Stellar Identifications (CSI) described by Ochsenbein is commended by Heck as a basic and reliable reference for difficult identifications as are frequently needed in operating the IUE satellite. The faint end ($V > 13$) of the CSI has a very patchy coverage of the sky, as stars are added from individual papers rather than from faint star surveys. This fact will need caution for some tasks, e.g., finding charts (Carnochan).

H. DATA IN ASTRONOMY

The discussion following the Invited Paper by Hauck was mainly concerned with the suggested filing system for photographic archives. Vette considered the informational aspects of the proposal quite tractable and analogous to space science procedures (observatories corresponding to spacecraft, telescope and accessories to missions, and plates to data sets). Albrecht described a set of computer programs at Vienna Observatory for plate cataloguing. The master file holds all the archival data, up to 30 parameters, including the few basic data Hauck had mentioned, but also plate scale, air mass, and others. An interactive data entry program and a printout utility allow to generate lists according to specific parameters (objects, coordinate limits, etc). Programs for sorting and for more sophisticated searches are expected to be added later. The archival system is written in Fortran IV; documentation and programs are available to interested persons upon request. A master file maintained at a data center, Albrecht continued, should certainly be as simple as possible for reasons of storage as well as file maintenance. In is quite unfeasible to request the maintenance of all conceivable data on a data center file. Once a researcher has located the potentially useful plates, more detailed information can be obtained by contacting the local plate archive system. Hauck announced that further consultations will take place on which items of information are required, and which others formatted but optional (as not all archives may be able to readily supply them). The preferred coordinates are α, δ 1950.0. The record should show if a plate is out on loan at the time of a second request. Since most request concern a certain object (star), priority attaches to developing computer lists of plates by object. For spectral plates, McCarthy suggests, the dispersion should be included with the basic sorting parameters (α, δ, name, m_V, date/time, plate number). Wilkins mentioned the problem of keeping photographic plates over long periods of time. Royal Greenwich Observatory would like to hear about work or techniques for extending the lifetime of plates, and for copying them economically while ensuring that their information contents is not degraded. McCarthy suggested consultations with curators of old plate collections, e.g., Carte du Ciel files. A mysterious malady called Gold Mold has recently been discovered, to which the important Kodak type III emulsions are subject. It may be related to changes in the fixing solutions.

Copying Schmidt plates without impairing quality was also addressed by Tritton. McCarthy commended the generosity in sharing the research materials with the whole astronomical community. Good positional precision seems to have been obtained from glass copies of ESO plates. The 8-parameter filing system of Benacchio was criticised by Albrecht who pointed out that many more parameters were advisable, so that the vital data are at hand without the need for frequent recourse to other sources such as observing logs.

The plates used in the study by Janes were Kodak IIaO and IIaD from a KPNO 91 cm telescope, scanned with a PDS at Yale. The accuracy obtained was encouraging; McCarthy referred to a recent paper by R. Albrecht (Ric.Astr.Spec.Vat. vol.10, p.9, 1980).

The suggestion by Egret and Jaschek, that the astronomical community should encourage the preparation of good catalogs and compilations, prompted this comment by Spite: There are prizes for papers; why not create an award, even symbolic, for the best catalog or compilation of the year? Carnochan mentioned that, with present videodisks and optical disks, very long term storage is possible by permanently burning information into the surface by laser; one such system, using computer-connected videodisks is already operational in the United States. Vette recalled a study by the Information Processing Division at Goddard Space Flight Center on the lifetime of magnetic tape storage, giving a much better result than many people have been led to believe: About one bit error per tape per year developed.

The new IDS catalog (Worley) will be in tape version only, and should be available through CDS and other centers in 1982. McCarthy mentioned new spectroscopic results and techniques (Michigan Survey, SIT Vidicon and Reticon as receivers in slit spectrographs). The catalog of clusters of galaxies (Nottale) has not yet advanced beyond the compilation of literature, and the circulation of a first draft is at least one year away.

The coordinate accuracy of extragalactic objects in available general catalogs is good enough for safe identifications, according to Paturel, so this material could be the nucleus for the central file. Cross-indexing with other object codes (for instances, radio sources) is not considered at this point.

The current catalog of galaxies (Kogoshvili) is in a computer form not adaptable to foreign computers; the new version now in preparation will be readable on IBM 360-line (800 bpi, 9 tracks, used code ABCDF), and should be ready for distribution by the end of 1981.

Jahreiss had this data breakdown for the 4597 stars which are candidates for inclusion in the FK5:

176 stars (4%) have trigonometric parallaxes determined,
937 stars (20%) have UBV photometry,
837 stars (18%) have MK spectral types,
1035 stars (23%) have radial velocities.

Basically there are records which software is available and where, though the detailed program documentation for outside use often is not (Grosbol). Computerisation problems may have already come up, and been solved, in different applications (as discussed between Lyngå and Davis in connection with open-cluster data); so it should pay to look around for potentially useful software. Very tough problems to tackle are storage of identification fields and cross-identification of extended objects with correspondingly uncertain positions.

I. DATA IN SPACE ASTRONOMY

NASA Stellar Data Center (NSSDC) has four periodic publications, as Vette summarized in response to Heck: The Report of Active and Planned Spacecraft and Experiments (RAPSE), the Launch Summary, and the NSSDC Data Listing are produced annually. RAPSE provides brief descriptions about spacecraft and experiments, about 1 paragraph in length, and identifies relevant personnel (Pricipal Investigators or Project Scientists). The Launch Summary includes all satellites and rockets in a given year; for the rocket launches the type of experiment and name of investigator are stated. The Data Listing is a one-line catalog of all NSSDC holdings, including ground-based data, environment models, and computer codes. The other periodic report is the monthly SPACEWARN Bulletin that provides information about recently launched or decayed satellites, as well as beacon frequencies and other parameters for satellites which are tracked optically. In addition, reports of studies, bibliographies, program summaries, and data books are published at various times. Some of the publications described by Mead and Warren at this Colloquium appear as NSSDC/WDC-A-R&S documents. Other activities at the NSSDC include a Satellite Situation Center, an Energetic Particle Environment Assessment Office, and recently the Science Operation Planner function for the Dynamics Explorer mission has been taken on. All of these activities include the preparation of data in special ways for individual missions.

Since packet telemetry is likely to be adopted for scientific satellites in the foreseeable future (de Pablo), packets and messages will have to be organized for the scientists and common elements arbitrated. Current plans at Goddard Space Flight Center also give appropriate attention to the archving aspect. Hauck was interested to learn about the fraction of costs of the NSSDC documentation and dissemination of satellite data for science in terms of the total mission expenses. Vette estimated that in the case of NASA-funded science experiments about 5% to 10% of the total cost go into the analysis of the data.

NSSDC represents about 2% of the funds used by NASA to support data analysis. In some cases it also serves as an archive for data from cooperative satellite missions with other countries, and for some experiments funded by other US government agencies.

GENERAL DISCUSSION

Among the speakers summarizing their impression of the overall significance of the Colloquium, Mead noted the progress made in the wide and expedient availability of catalogued data. The 1976 conference had brought together many astronomical experts with problems and efforts in their respective subject areas, while presently the growing participation of data-handling experts provided increased know-how toward solutions. Heck emphasized the necessity to update catalogued data with respect to errors detected. Corrections should be published in cumulative form, and - in doubtful cases - the catalog version stored or used will need to be specified. He also expect more interfacing use between space and groundbased astronomical data. Worley addressed the problem of identifiers; he warned against proliferation of designations, recommended universally used identifiers, and defended the right of the IAU to endorse such identifiers as requisite for expedient documentation.

The wide ranging scope of this Colloquium, from software and datacenter operation to abstracts and patents, was appreciated in the summary by Heintz. The Working Groups in IAU Commission 5 have service and "watchdog" functions in these areas, and the coordination toward which they work should help counteract disintegration of subject areas within the IAU as well as cross-discipline. The Guidelines presented by Wilkins and other coordinative documents mentioned at this conference deserve wide attention; Kleczek is putting final touches on the revision of the Astronomical Dictionary, and the IAU is represented in ICSU, CODATA, the Abstracting Board, and other organizations. Problem areas are the data flux in view of hardware incompatibilities, divergent needs in data documentation between subject areas, the pros and cons of data publication versus storage, and the search for optimum cost-efficiency in data management which includes careful assessment of allowable redundancy.

Jaschek and others mentioned the use of positional catalogs as identifier sources (inventory lists). SAOC and AGK3 are most demanded. For this purpose the completeness of SAOC leaves something to be desired and a replacement should be considered, in view of the large amount of work already invested in positions and proper motions of fainter stars, and of the plate material still being measured. Inclusion of bibliographical references in materials for data centers is certainly desired. In closing, Hauck expressed satisfaction over the increasingly useful function of data centers and networks. The skepticism of a few years back about their feasibility in astronomy has long been replaced by appreciation of the success of an activity, the growth of which is far from ended.

AUTHOR'S INDEX

ASTROPHYSICS AND SPACE SCIENCE LIBRARY

Edited by

J. E. Blamont, R. L. F. Boyd, L. Goldberg, C. de Jager, Z. Kopal, G. H. Ludwig, R. Lüst, B. M. McCormac, H. E. Newell, L. I. Sedov, Z. Švestka, and W. de Graaff

1. C. de Jager (ed.), *The Solar Spectrum, Proceedings of the Symposium held at the University of Utrecht, 26–31 August, 1963.* 1965, XIV + 417 pp.
2. J. Orthner and H. Maseland (eds.), *Introduction to Solar Terrestrial Relations, Proceedings of the Summer School in Space Physics held in Alpbach, Austria, July 15–August 10, 1963 and Organized by the European Preparatory Commission for Space Research.* 1965, IX + 506 pp.
3. C. C. Chang and S. S. Huang (eds.), *Proceedings of the Plasma Space Science Symposium, held at the Catholic University of America, Washington, D.C., June 11–14, 1963.* 1965, IX + 377 pp.
4. Zdeněk Kopal, *An Introduction to the Study of the Moon.* 1966, XII + 464 pp.
5. B. M. McCormac (ed.), *Radiation Trapped in the Earth's Magnetic Field. Proceedings of the Advanced Study Institute, held at the Chr. Michelsen Institute, Bergen, Norway, August 16–September 3, 1965.* 1966, XII + 901 pp.
6. A. B. Underhill, *The Early Type Stars.* 1966, XII + 282 pp.
7. Jean Kovalevsky, *Introduction to Celestial Mechanics.* 1967, VIII + 427 pp.
8. Zdeněk Kopal and Constantine L. Goudas (eds.), *Measure of the Moon. Proceedings of the 2nd International Conference on Selenodesy and Lunar Topography, held in the University of Manchester, England, May 30–June 4, 1966.* 1967, XVIII + 479 pp.
9. J. G. Emming (ed.), *Electromagnetic Radiation in Space. Proceedings of the 3rd ESRO Summer School in Space Physics, held in Alpbach, Austria, from 19 July to 13 August, 1965.* 1968, VIII + 307 pp.
10. R. L. Carovillano, John F. McClay, and Henry R. Radoski (eds.), *Physics of the Magnetosphere, Based upon the Proceedings of the Conference held at Boston College, June 19–28, 1967.* 1968, X + 686 pp.
11. Syun-Ichi Akasofu, *Polar and Magnetospheric Substorms.* 1968, XVIII + 280 pp.
12. Peter M. Millman (ed.), *Meteorite Research. Proceedings of a Symposium on Meteorite Research, held in Vienna, Austria, 7–13 August, 1968.* 1969, XV + 941 pp.
13. Margherita Hack (ed.), *Mass Loss from Stars. Proceedings of the 2nd Trieste Colloquium on Astrophysics, 12–17 September, 1968.* 1969, XII + 345 pp.
14. N. D'Angelo (ed.), *Low-Frequency Waves and Irregularities in the Ionosphere. Proceedings of the 2nd ESRIN-ESLAB Symposium, held in Frascati, Italy, 23–27 September, 1968.* 1969, VII + 218 pp.
15. G. A. Partel (ed.), *Space Engineering. Proceedings of the 2nd International Conference on Space Engineering, held at the Fondazione Giorgio Cini, Isola di San Giorgio, Venice, Italy, May 7–10, 1969.* 1970, XI + 728 pp.
16. S. Fred Singer (ed.), *Manned Laboratories in Space. Second International Orbital Laboratory Symposium.* 1969, XIII + 133 pp.
17. B. M. McCormac (ed.), *Particles and Fields in the Magnetosphere. Symposium Organized by the Summer Advanced Study Institute, held at the University of California, Santa Barbara, Calif., August 4–15, 1969.* 1970, XI + 450 pp.
18. Jean-Claude Pecker, *Experimental Astronomy.* 1970, X + 105 pp.
19. V. Manno and D. E. Page (eds.), *Intercorrelated Satellite Observations related to Solar Events. Proceedings of the 3rd ESLAB/ESRIN Symposium held in Noordwijk, The Netherlands, September 16–19, 1969.* 1970, XVI + 627 pp.
20. L. Mansinha, D. E. Smylie, and A. E. Beck, *Earthquake Displacement Fields and the Rotation of the Earth, A NATO Advanced Study Institute Conference Organized by the Department of Geophysics, University of Western Ontario, London, Canada, June 22–28, 1969.* 1970, XI + 308 pp.
21. Jean-Claude Pecker, *Space Observatories.* 1970, XI + 120 pp.
22. L. N. Mavridis (ed.), *Structure and Evolution of the Galaxy. Proceedings of the NATO Advanced Study Institute, held in Athens, September 8–19, 1969.* 1971, VII + 312 pp.

23. A. Muller (ed.), *The Magellanic Clouds. A European Southern Observatory Presentation: Principal Prospects, Current Observational and Theoretical Approaches, and Prospects for Future Research, Based on the Symposium on the Magellanic Clouds, held in Santiago de Chile, March 1969, on the Occasion of the Dedication of the European Southern Observatory.* 1971, XII + 189 pp.
24. B. M. McCormac (ed.), *The Radiating Atmosphere. Proceedings of a Symposium Organized by the Summer Advanced Study Institute, held at Queen's University, Kingston, Ontario, August 3–14, 1970.* 1971, XI + 455 pp.
25. G. Fiocco (ed.), *Mesospheric Models and Related Experiments. Proceedings of the 4th ESRIN-ESLAB Symposium, held at Frascati, Italy, July 6–10, 1970.* 1971, VIII + 298 pp.
26. I. Atanasijević, *Selected Exercises in Galactic Astronomy.* 1971, XII + 144 pp.
27. C. J. Macris (ed.), *Physics of the Solar Corona. Proceedings of the NATO Advanced Study Institute on Physics of the Solar Corona, held at Cavouri-Vouliagmeni, Athens, Greece, 6–17 September 1970.* 1971, XII + 345 pp.
28. F. Delobeau, *The Environment of the Earth.* 1971, IX + 113 pp.
29. E. R. Dyer (general ed.), *Solar-Terrestrial Physics/1970. Proceedings of the International Symposium on Solar-Terrestrial Physics, held in Leningrad, U.S.S.R., 12–19 May 1970.* 1972, VIII + 938 pp.
30. V. Manno and J. Ring (eds.), *Infrared Detection Techniques for Space Research. Proceedings of the 5th ESLAB-ESRIN Symposium, held in Noordwijk, The Netherlands, June 8–11, 1971.* 1972, XII + 344 pp.
31. M. Lecar (ed.), *Gravitational N-Body Problem. Proceedings of IAU Colloquium No. 10, held in Cambridge, England, August 12–15, 1970.* 1972, XI + 441 pp.
32. B. M. McCormac (ed.), *Earth's Magnetospheric Processes. Proceedings of a Symposium Organized by the Summer Advanced Study Institute and Ninth ESRO Summer School, held in Cortina, Italy, August 30–September 10, 1971.* 1972, VIII + 417 pp.
33. Antonin Rükl, *Maps of Lunar Hemispheres.* 1972, V + 24 pp.
34. V. Kourganoff, *Introduction to the Physics of Stellar Interiors.* 1973, XI + 115 pp.
35. B. M. McCormac (ed.), *Physics and Chemistry of Upper Atmospheres. Proceedings of a Symposium Organized by the Summer Advanced Study Institute, held at the University of Orléans, France, July 31–August 11, 1972.* 1973, VIII + 389 pp.
36. J. D. Fernie (ed.), *Variable Stars in Globular Clusters and in Related Systems. Proceedings of the IAU Colloquium No. 21, held at the University of Toronto, Toronto, Canada, August 29–31, 1972.* 1973, IX + 234 pp.
37. R. J. L. Grard (ed.), *Photon and Particle Interaction with Surfaces in Space. Proceedings of the 6th ESLAB Symposium, held at Noordwijk, The Netherlands, 26–29 September, 1972.* 1973, XV + 577 pp.
38. Werner Israel (ed.), *Relativity, Astrophysics and Cosmology. Proceedings of the Summer School, held 14–26 August, 1972, at the BANFF Centre, BANFF, Alberta, Canada.* 1973, IX + 323 pp.
39. B. D. Tapley and V. Szebehely (eds.), *Recent Advances in Dynamical Astronomy. Proceedings of the NATO Advanced Study Institute in Dynamical Astronomy, held in Cortina d'Ampezzo, Italy, August 9–12, 1972.* 1973, XIII + 468 pp.
40. A. G. W. Cameron (ed.), *Cosmochemistry. Proceedings of the Symposium on Cosmochemistry, held at the Smithsonian Astrophysical Observatory, Cambridge, Mass., August 14–16, 1972.* 1973, X + 173 pp.
41. M. Golay, *Introduction to Astronomical Photometry.* 1974, IX + 364 pp.
42. D. E. Page (ed.), *Correlated Interplanetary and Magnetospheric Observations. Proceedings of the 7th ESLAB Symposium, held at Saulgau, W. Germany, 22–25 May, 1973.* 1974, XIV + 662 pp.
43. Riccardo Giacconi and Herbert Gursky (eds.), *X-Ray Astronomy.* 1974, X + 450 pp.
44. B. M. McCormac (ed.), *Magnetospheric Physics. Proceedings of the Advanced Summer Institute, held in Sheffield, U.K., August 1973.* 1974, VII + 399 pp.
45. C. B. Cosmovici (ed.), *Supernovae and Supernova Remnants. Proceedings of the International Conference on Supernovae, held in Lecce, Italy, May 7–11, 1973.* 1974, XVII + 387 pp.
46. A. P. Mitra, *Ionospheric Effects of Solar Flares.* 1974, XI + 294 pp.
47. S.-I. Akasofu, *Physics of Magnetospheric Substorms.* 1977, XVIII + 599 pp.

48. H. Gursky and R. Ruffini (eds.), *Neutron Stars, Black Holes and Binary X-Ray Sources.* 1975, XII + 441 pp.
49. Z. Švestka and P. Simon (eds.), *Catalog of Solar Particle Events 1955–1969. Prepared under the Auspices of Working Group 2 of the Inter-Union Commission on Solar-Terrestrial Physics.* 1975, IX + 428 pp.
50. Zdeněk Kopal and Robert W. Carder, *Mapping of the Moon.* 1974, VIII + 237 pp.
51. B. M. McCormac (ed.), *Atmospheres of Earth and the Planets. Proceedings of the Summer Advanced Study Institute, held at the University of Liège, Belgium, July 29–August 8, 1974.* 1975, VII + 454 pp.
52. V. Formisano (ed.), *The Magnetospheres of the Earth and Jupiter. Proceedings of the Neil Brice Memorial Symposium, held in Frascati, May 28–June 1, 1974.* 1975, XI + 485 pp.
53. R. Grant Athay, *The Solar Chromosphere and Corona: Quiet Sun.* 1976, XI + 504 pp.
54. C. de Jager and H. Nieuwenhuijzen (eds.), *Image Processing Techniques in Astronomy. Proceedings of a Conference, held in Utrecht on March 25–27, 1975.* XI + 418 pp.
55. N. C. Wickramasinghe and D. J. Morgan (eds.), *Solid State Astrophysics. Proceedings of a Symposium, held at the University College, Cardiff, Wales, 9–12 July 1974.* 1976, XII + 314 pp.
56. John Meaburn, *Detection and Spectrometry of Faint Light.* 1976, IX + 270 pp.
57. K. Knott and B. Battrick (eds.), *The Scientific Satellite Programme during the International Magnetospheric Study. Proceedings of the 10th ESLAB Symposium, held at Vienna, Austria, 10–13 June 1975.* 1976, XV + 464 pp.
58. B. M. McCormac (ed.), *Magnetospheric Particles and Fields. Proceedings of the Summer Advanced Study School, held in Graz, Austria, August 4–15, 1975.* 1976, VII + 331 pp.
59. B. S. P. Shen and M. Merker (eds.), *Spallation Nuclear Reactions and Their Applications.* 1976, VIII + 235 pp.
60. Walter S. Fitch (ed.), *Multiple Periodic Variable Stars. Proceedings of the International Astronomical Union Colloquium No. 29, held at Budapest, Hungary, 1–5 September 1976.* 1976, XIV + 348 pp.
61. J. J. Burger, A. Pedersen, and B. Battrick (eds.), *Atmospheric Physics from Spacelab. Proceedings of the 11th ESLAB Symposium, Organized by the Space Science Department of the European Space Agency, held at Frascati, Italy, 11–14 May 1976.* 1976, XX + 409 pp.
62. J. Derral Mulholland (ed.), *Scientific Applications of Lunar Laser Ranging. Proceedings of a Symposium held in Austin, Tex., U.S.A., 8–10 June, 1976.* 1977, XVII + 302 pp.
63. Giovanni G. Fazio (ed.), *Infrared and Submillimeter Astronomy. Proceedings of a Symposium held in Philadelphia, Penn., U.S.A., 8–10 June, 1976.* 1977, X + 226 pp.
64. C. Jaschek and G. A. Wilkins (eds.), *Compilation, Critical Evaluation and Distribution of Stellar Data. Proceedings of the International Astronomical Union Colloquium No. 35, held at Strasbourg, France, 19–21 August, 1976.* 1977, XIV + 316 pp.
65. M. Friedjung (ed.), *Novae and Related Stars. Proceedings of an International Conference held by the Institut d'Astrophysique, Paris, France, 7–9 September, 1976.* 1977, XIV + 228 pp.
66. David N. Schramm (ed.), *Supernovae. Proceedings of a Special IAU-Session on Supernovae held in Grenoble, France, 1 September, 1976.* 1977, X + 192 pp.
67. Jean Audouze (ed.), *CNO Isotopes in Astrophysics. Proceedings of a Special IAU Session held in Grenoble, France, 30 August, 1976.* 1977, XIII + 195 pp.
68. Z. Kopal, *Dynamics of Close Binary Systems,* XIII + 510 pp.
69. A. Bruzek and C. J. Durrant (eds.), *Illustrated Glossary for Solar and Solar-Terrestrial Physics.* 1977, XVIII + 204 pp.
70. H. van Woerden (ed.), *Topics in Interstellar Matter.* 1977, VIII + 295 pp.
71. M. A. Shea, D. F. Smart, and T. S. Wu (eds.), *Study of Travelling Interplanetary Phenomena.* 1977, XII + 439 pp.
72. V. Szebehely (ed.), *Dynamics of Planets and Satellites and Theories of Their Motion. Proceedings of IAU Colloquium No. 41, held in Cambridge, England, 17–19 August 1976.* 1978, XII + 375 pp.
73. James R. Wertz (ed.), *Spacecraft Attitude Determination and Control.* 1978, XVI + 858 pp.

74. Peter J. Palmadesso and K. Papadopoulos (eds.), *Wave Instabilities in Space Plasmas. Proceedings of a Symposium Organized Within the XIX URSI General Assembly held in Helsinki, Finland, July 31–August 8, 1978.* 1979, VII + 309 pp.
75. Bengt E. Westerlund (ed.), *Stars and Star Systems. Proceedings of the Fourth European Regional Meeting in Astronomy held in Uppsala, Sweden, 7–12 August, 1978.* 1979, XVIII + 264 pp.
76. Cornelis van Schooneveld (ed.), *Image Formation from Coherence Functions in Astronomy. Proceedings of IAU Colloquium No. 49 on the Formation of Images from Spatial Coherence Functions in Astronomy, held at Groningen, The Netherlands, 10–12 August 1978.* 1979, XII + 338 pp.
77. Zdeněk Kopal, *Language of the Stars. A Discourse on the Theory of the Light Changes of Eclipsing Variables.* 1979, VIII + 280 pp.
78. S.-I. Akasofu (ed.), *Dynamics of the Magnetosphere. Proceedings of the A.G.U. Chapman Conference 'Magnetospheric Substorms and Related Plasma Processes' held at Los Alamos Scientific Laboratory, N.M., U.S.A., October 9-13, 1978.* 1980, XII + 658 pp.
79. Paul S. Wesson, *Gravity, Particles, and Astrophysics. A Review of Modern Theories of Gravity and G-variability, and their Relation to Elementary Particle Physics and Astrophysics.* 1980, VIII + 188 pp.
80. Peter A. Shaver (ed.), *Radio Recombination Lines. Proceedings of a Workshop held in Ottawa, Ontario, Canada, August 24-25, 1979.* 1980, X + 284 pp.
81. Pier Luigi Bernacca and Remo Ruffini (eds.), *Astrophysics from Spacelab*, 1980, XI + 664 pp.
82. Hannes Alfvén, *Cosmic Plasma*, 1981, X + 160 pp.
83. Michael D. Papagiannis (ed.), *Strategies for the Search for Life in the Universe*, 1980, XVI + 254 pp.
84. H. Kikuchi (ed.), *Relation between Laboratory and Space Plasmas*, 1981, XII + 386 pp.
85. Peter van der Kamp, *Stellar Paths*, 1981, xxii + 155 pp.
86. E. M. Gaposchkin and B. Kołaczek (eds.), *Reference Coordinate Systems for Earth Dynamics*, 1981, XIV + 396 pp.
87. R. Giacconi (ed.), *X-Ray Astronomy with the Einstein Satellite. Proceedings of the High Energy Astrophysics Division of the American Astronomical Society Meeting on X-Ray Astronomy held at the Harvard-Smithsonian Center for Astrophysics, Cambridge, Mass., U.S.A., January 28–30, 1980.* 1981, VII + 330 pp.
88. Icko Iben Jr. and Alvio Renzini (eds.), *Physical Processes in Red Giants. Proceedings of the Second Workshop, held at the Ettore Majorana Centre for Scientific Culture, Advanced School of Agronomy, in Erice, Sicily, Italy, September 3–13, 1980.* 1981, XV + 488 pp.
89. C. Chiosi and R. Stalio (eds.), *Effect of Mass Loss on Stellar Evolution. IAU Colloquium No. 59 held in Miramare, Trieste, Italy, September 15–19, 1980.* 1981, XXII + 532 pp.
90. C. Goudis, *The Orion Complex: A Case Study of Interstellar Matter*, 1982 (forthcoming).
91. F. D. Kahn (ed.), *Investigating the Universe. Papers Presented to Zdeněk Kopal on the Occasion of his retirement, September 1981.* 1981, X + 458 pp.
92. C. M. Humphries (ed.), *Instrumentation for Astronomy with Large Optical Telescopes, Proceedings of IAU Colloquium No. 67.* 1982 (forthcoming).
93. R. S. Roger and P. E. Dewdney (eds.), *Regions of Recent Star Formation, Proceedings of the Symposium on "Neutral Clouds Near HII Regions - Dynamics and Photochemistry", held in Penticton, B.C., June 24–26, 1981.* 1982, XVI + 496 pp.
94. O. Calame (ed.), *High-Precision Earth Rotation and Earth-Moon Dynamics. Lunar Distances and Related Observations*, 1982, xx + 354 pp.
95. M. Friedjung and R. Viotti (eds.), *The Nature of Symbiotic Stars*, xx + 310 pp.
96. W. Fricke and G. Teleki (eds.), *Sun and Planetary System*, xiv + 538 pp.
97. C. Jaschek and W. Heintz (eds.), *Automated Data Retrieval in Astronomy*, xx + 324 pp.

UNIVERSITY COLLEGE
LIBRARY
CARDIFF